D0908655

BUSINESS ASPECTS OF
TECHNOLOGY TRANSFER

BUSINESS ASPECTS OF TECHNOLOGY TRANSFER

Marketing and Acquisition

by

William M. Watkins

np NOYES PUBLICATIONS
Park Ridge, New Jersey, U.S.A.

658.18

Library of Congress Catalog Card Number: 89-8736
ISBN: 0-8155-1206-6
Printed in the United States

Published in the United States of America by
Noyes Publications
Mill Road, Park Ridge, New Jersey 07656

10 9 8 7 6 5 4 3 2 1

Library of Congress Cataloging-in-Publication Data

Watkins, William M.
Business aspects of technology transfer : marketing and
acquisition / by William M. Watkins.
p.   cm.
ISBN 0-8155-1206-6 :
1. Technology transfer--Management. 2. Foreign licensing
agreements--Management. I. Title.
HD45.W39   1989
658.1'8--dc20   89-8736
                                        CIP

To Lura

Preface

It is probably fair to say that every author sees a need for his book, and this particular effort is no exception. One of the unique characteristics of the technology transfer or licensing field is the fact that many of its practitioners have had to undergo "on-the-job training" because of the lack of opportunities for formal academic preparation for a career in licensing. Except for attorneys, most enter the field of technology transfer from a background in engineering, science, or related commercial areas and proceed to learn by doing. Although there is little doubt that such a practical approach can be an effective form of training and growth, there is also a strong belief that an understandable, easily-accessed reference book can facilitate the transition from other fields into a career in technology transfer. In addition, it is hoped that such a resource can also provide a useful overview of the field of technology transfer for those managers and professionals who must deal with various aspects of the area.

It is this consideration that has led to the title *Business Aspects of Technology Transfer*. All too often, technology transfer is overlooked or neglected as an effective element of business strategy. In the belief that technology transfer, or licensing, should be treated as a viable business option, this book has been structured to reflect those activities and functions that are common to most business ventures. These include the familiar elements of market research, packaging, pricing, promotion and advertising, sales, and support activities. In addition, the fact that technology transfer is a two-way process is reflected by the division of the book into two principal sections: technology marketing and technology acquisition.

Finally, an attempt has been made to condense the treatment of these various activities into an understandable overview suitable for those seeking a working familiarity with the area. The primary objective has been breadth of coverage rather than an in-depth treatment of a few specific components. Above all, the hope is that this effort will contribute, in some small way, to a better understanding and appreciation of technology transfer as an effective business option.

W.M. Watkins

About the Author

William M. Watkins, Manager of Process Engineering for MTI Corporation, St. Albans, West Virginia, and formerly Manager of Licensing for Union Carbide's Process Systems Department, has been involved in licensing activities in over 30 countries around the world. He is currently serving on the Latin American and Small Business committees of the Licensing Executives Society, and is also serving as a consultant to a Fortune 500 company on the licensing of technology to a major Japanese firm. In addition, he is active as a consultant to small businesses under the Small Business Administration.

NOTICE

To the best of the Publisher's knowledge the information contained in this book is accurate; however, the Publisher assumes no responsibility nor liability for errors or any consequences arising from the use of the information contained herein. Final determination of the suitability of any information for use contemplated by any user, and the manner of that use, is the sole responsibility of the user.

Contents and Subject Index

PART III
TECHNOLOGY ACQUISITION AND IN-LICENSING

PART IV
APPENDIX

Part I

The Technology Product

1

An Overview of
Technology Transfer

Many corporate managers and technical personnel have been introduced to technology transfer on a "trial-by-fire" or "learn-by-doing" basis. Although knowledgeable and experienced in their particular functions, their first contact with technology transfer may be a key assignment to manage or execute the marketing/transfer of some portion of their organization's knowledge. A decision has been made, perhaps as the result of an external inquiry, to attempt to license a technology. A proven performer in marketing plastics, metallurgical products, or chemicals will find that marketing technology, a very intangible product, can be a completely different challenge with new problems, issues, and pitfalls. A scientist or engineer, used to dealing with technology as simply a means to an end, suddenly finds that the technology is now the product itself. Whether trained in the liberal arts or the sciences, one usually finds little in his or her academic training that will serve as preparation for an assignment in licensing or technology transfer. The concept of promoting, selling, and delivering a product that, except for its packaging, cannot be touched or measured in tangible terms is a very different, and sometimes unsettling, experience for a manager or scientist. One of the purposes of this book is to provide a practical resource in technology transfer for managerial or technical personnel in either industrial or academic fields.

There is an even more compelling reason for another book on technology transfer. This is the belief that technology is probably the most financially underutilized asset in either the private or public sector and in almost any developed nation of the world. There is a continuing effort to ensure effective utilization and a decent financial yield on tangible assets, such as manufacturing facilities, rolling stock, and real estate. Yet these same organizations may possess technology and know-how that is either not used or used only for captive applications.

It may be obvious, but bears repeating, that technological assets have some unique and very attractive qualities not always available in other assets. First, technological products in the form of licenses or rights to use the technology can be sold repeatedly with a relatively low "cost-of-goods-sold." Successful management of a technology transfer program can produce repeated streams of income, while at the same time preserving or even enhancing the value of the original asset. Second, technological products represent one of the few articles of commerce for which governments will provide and sanction monopoly rights for an extended period of time in the form of patent coverage. Another objective of this book, therefore, is to provide additional know-how for the non-licensing professional to capitalize on these "hidden assets."

Given these objectives, the primary effort has been to provide a practical, easy-to-access source of technology transfer know-how. It is assumed that the reader is knowledgeable in his or her field, but has not had extensive experience in technology transfer. The emphasis, therefore, is on the "how-to" aspects of licensing. Accordingly, the book is organized to address such specific areas as:

o   Deciding to market technology

o   Technology market research

o   Product packaging

o   Product pricing

o   Promotion and advertising

o   Technology transfer agreements

o   Service and support

o   Technology acquisition

o   Technology searches

o   Technology evaluation

Some preparation is required, however, before addressing these more definitive areas. It is helpful, first, to develop a basis for understanding by briefly considering some of the continuing issues and historical trends in technology transfer. Second, and probably even more important to understanding, is having a clear picture of the nature of intellectual property, in general, and of technology as a product, in particular. There are, indeed, significant differences in the management and exploitation of technological assets as opposed to the more widely understood fixed assets. For this reason, a general discussion of technology as a product is included in Chapter 2 to set the stage for the more definitive discussions to come.

ISSUES IN TECHNOLOGY TRANSFER

Technology transfer can be an extraordinarily lucrative enterprise, even with the significant effort and management required. Such technological assets as the Monsanto acetic acid technology, the Union Carbide polyolefins technology, and the Bayer vinyl acetate process have yielded hundreds of millions of dollars for their owners over the years. However, technology transfer is not always the "right" answer. Certain difficult issues must be addressed and resolved in terms of the owner's unique situation and objectives.

A basic issue in almost any technology transfer, both for the licensor and the licensee, is that of technology development versus technology transfer, or more simply stated, the "make-versus-buy" question. An individual or organization always has the option of developing or "inventing" a technology, as opposed to paying what may even be a competitor. The question is a difficult one, to which there is no easy answer. Yet it is this issue, present in nearly every technology need, that determines marketability, value, and competitiveness.

A second issue is nearly always present in the decision to market or license one's technology and know-how. This is the role of technology transfer in a manufacturer's overall business strategy. To paraphrase, *"Should we use our valuable self-developed technology to achieve a competitive advantage in our own manufacturing costs, or should we actively market the right to use our technology as a co-product of the commodity itself? If the latter, how do we allocate our management and marketing efforts between the technology product and the tangible product to optimize our total yield on the effort?"* Again, not an easy question, but rather one that can hold its own with religion and politics as a source of emotional trauma.

A third, and somewhat related, issue is the political counterpart to the commercial issues outlined above. There is increasing recognition throughout the world that technology and its transfer can be a potent instrument of national policy. For many nations, access to weapons technology, commercial aircraft hardware and know-how, and one's own basic steel industry have been the cornerstones of government stability (or lack thereof). The issue here is how an industrial country can use its technological assets to best achieve its diplomatic and trade objectives.

There are, obviously, no universal solutions or pat answers to these difficult questions. Time, circumstances, culture, noble aspiration, and pure greed all affect the choice of a course of action. However, a look back at technology transfer, together with a recognition of current trends, can provide some perspective.

HISTORICAL TRENDS

The history of technology transfer in the world is best left to those with a strong historical bent. Needless to say, technologies such as fire-building and the wheel were probably the subjects of de facto license agreements. Some far more recent trends in technology transfer, however, can provide some useful perspective on the issues previously discussed.

The first of these is the transition from a trade secret mentality, a.k.a. "The Crown Jewel Syndrome," to the acceptance of technology transfer as a respectable element of business strategy. A prevalent belief in the 1920s and 1930s was that technology was most valuable when carefully guarded and applied only to captive applications. In many instances, patent protection was shunned because of the publication requirements. With greater personnel mobility and its impact on trade secret preservation, the treatment of technology as an article of commerce has grown. More importantly, there is increasing recognition that the yield on crown jewels, or trade secrets, is often much less than that of a well-managed licensing program.

A second, and related, trend involves the willingness to apply marketing strengths and resources to technology transfer efforts. Prior to World War II, and even thereafter, the total licensing program in many organizations could be paraphrased as "We're willing to talk, if someone's interested in our technology." Sometimes known as "The Better Mousetrap Syndrome," this outlook characterized technology transfer, at best, as an annoying distraction and, at worst, as a threat to corporate security. Fortunately, we now see many well-managed, vigorous licensing organizations, and the trend is clearly one of recognizing technology transfer as an effective instrument of corporate strategy.

Looking to the future, there are some notable trends in technology transfer. Technology transfer will continue to be more globally oriented. Barriers to free movement of technology across national borders will exist, but will be modified as necessary to fit national goals. Thus, in many parts of the world, government's role in the transactions involving intellectual property will increase in significance.

In the private sector, there will be increasing recognition that manufactured items, be they electronic, chemical, or other, can be managed as co-products of the technology required to manufacture them, and the yield on both types of products can thus be maximized.

In summary, technology assets are intangible and often unrecognized and underutilized. A clear understanding of the techniques and approaches to technology transfer can serve to increase significantly the financial yield on these neglected assets. Captive use of technology and an active program of technology transfer can co-exist, not only peacefully, but quite profitably. The need for individuals with a practical understanding of technology transfer, both as licensors and licensees, will continue to grow steadily.

In the following chapters, we will be looking at the selection and identification of technology products, market definition, and the packaging of the products for sale. In addition, we will examine the broad area of pricing and valuation of technological products, as well as related financial considerations. Promotion, advertising, and sales efforts will also be reviewed as part of the licensor perspective. The other side of the transaction will be examined in a section on the acquisition or purchase of technology.

Although this book is in no way a substitute for the services of an attorney, a section on the licensing agreement itself is presented from the point of view of an industrial executive. Finally, but by no means least, an overview of the services and support obligation associated with technology transfer is presented.

2

Technology As a Product

One of our Supreme Court associate justices is reputed to have said of pornography that he could not define it, but he was confident he could recognize it when he saw it. The same statement could be made with regard to technological products. We tend to think of property in terms of *what it is*. Real estate, for example, can be characterized as a particular location, as a specific size, as a certain geometric shape, and as flat or mountainous, to mention only a very few qualities. An automobile is defined in terms of size, color, power, styling, and a host of other tangible properties. Clothing, machinery, jewelry—all these and many others can be described by their intrinsic qualities, usually very visible and explicit.

True technological products, on the other hand, can not be described in the same way. A technological product is usually defined in terms of what it permits the user to do. This distinction between tangible property and technological property may appear to be subtle, but it is far-reaching in its consequences.

Consider, for example, a new type of catalyst. Let us assume that it is constructed of a new space-age composite material, lasts twice as long as other catalysts, and provides higher productivity. You are the product manager, so how do you go about the preparation of a product marketing plan and its implementation? The answer, of course, is that we are talking of not one, but *two separate products*. We have, on the one hand, a tangible physical product, the catalyst itself. For this product, there are specific target markets, prices and pricing philosophies to be established, advertising and sales promotion efforts to be defined, and a host of other issues to be resolved.

Note, however, that there is a second product that must be addressed. This is the technological product, or the know-how required to manufacture this catalyst. Someone in Research and Development (R&D) or Engineering, perhaps, has developed an idea or concept into a technological product that permits one to mass-produce a tangible physical product—the miracle catalyst. How does one describe or characterize this technological product? Most assuredly "with great difficulty," but most probably in terms of what it permits one to do, namely, produce the "World's Greatest Catalyst."

You have developed a marketing plan for the catalyst itself. You have identified target markets and volumes, outlined pricing strategy, characterized competitive processes and possible responses, and scoped out promotional efforts. Does this mean that you go through the same exercise for the technological product? The answer is a "definite maybe," depending on how you or your organization responds to questions, such as the following:

o   What does it take (money, people, time) to market (i.e., license) the technological product?

o   What is the impact on our catalyst sales of giving competitors access to the technology through licensing?

o   How is the technological product priced? What pricing format—lump sum, running royalty, or a combination?

o   Do we work with a third party, such as a contractor or equipment vendor, or do we work alone?

o   Globally, for each geographical territory, do we:

      (1)   market catalyst only?
      (2)   market catalyst technology only?
      (3)   market both?

o   How do we combine our ownership of these two products, catalyst and catalyst technology, to maximize our financial yield over the long term?

o   For our technology product, how do we disclose enough for a potential licensee to evaluate without giving away too much information?

o   What are the important issues to be addressed and resolved in a licensing agreement?

o   What technical resources will be required to sustain a successful licensing effort?

o   Culturally and organizationally, can technology licensing and product marketing/manufacturing peacefully co-exist in our organization?

These are a few of the technology product issues this book is intended to address. Before proceeding further, however, let us step back and take a very brief look at the broad category to which technology products belong, that of intellectual property.

INTELLECTUAL PROPERTY

In general, intellectual property consists of know-how, patents, copyrights, and trademarks. By and large, technology transfer as used in this book involves usually the first two—know-how and patents. However, for the sake of an orderly progression through these various categories of intellectual property, a brief look at each is appropriate.

Before doing so, however, a disclaimer is in order. This book is not intended as a do-it-yourself manual for the practice of patent, copyright, or trademark law. Neither is it designed to substitute for the professional involvement of an attorney in the drafting and interpretation of contractual arrangements in the area of technology transfer. Nevertheless, for anyone involved in either the commercial or technical aspects of technology transfer, an understanding of the various categories is highly desirable.

Know-How

Probably one of the more non-specific terms in intellectual property, know-how is involved in essentially every instance of technology transfer. The entire field of technology transfer is based on the premise of an individual or organization knowing how to do something and being willing to share that know-how with someone else. The first instance of know-how transfer is undoubtedly lost in antiquity; one can visualize, however, prehistoric techniques for survival, weapons development, and medical treatment as examples of know-how. Even in those days, before being known as a category of intellectual property, know-how had value. Knowing how to do something or make something, that others did not, conferred status, authority, and power. We shall not attempt to extend this by contending that the first licensing manager was a tribal witch doctor, but the idea of the value of know-how has been around for a long time.

Know-how, therefore, can be transferred, or licensed for use in return for money, goods, or other valuable consideration. It is important to note, however, that all know-how is not a candidate for licensing or commercial technology transfer. Certain requirements must be met by the know-how to make it a valuable article of commerce.

First, it must have at least the perception of value or usefulness in the eyes of the potential buyer. Regardless of the cleverness or elegance of the concept, the technological product, or know-how, must be, in most cases, something more than an intellectual exercise or new thought concept.

Second, in order to be of commercial utility, there must be limited access to the know-how. Know-how comes in many how-to manuals on nearly every conceivable subject. In this technology transfer context, potentially licensable know-how is that which has been developed or acquired by its owner and carefully limited in circulation and disclosure in order to enhance its commercial value.

Third, and perhaps most importantly, note that know-how can be transferred or licensed without being patented. Although patent coverage enhances the value of the know-how, not all valuable know-how is eligible for patent coverage. With that in mind, let us take a look at the second category of intellectual property, in reality a subclass of know-how in general—patents.

Patents

Certain types of know-how, meeting specific requirements, have been declared by governments to be eligible for the protection of a patent. A patent thus granted gives the recipient the right to exclude others from making, using, or selling the know-how in a particular geographical area. The intention, of course, is to provide an incentive for new discoveries in science and the useful arts. For a period of time, (17 years in the U.S.), a patent provides the developer of certain know-how a veritable monopoly in its implementation, use, and sale. Note, however, that the duration of the monopoly is limited in order to serve the ultimate objective, i.e., that of making the benefits of the know-how broadly available to society in general.

The law, however, carefully limits the kind of know-how that can be patented:

o   The know-how must be *new and useful.* An old, long-practiced technology is not patentable simply because no one else has coverage. By the same token, a scientific curiosity or novelty is not, of and by itself, patentable unless it has use.

o   The know-how may be in various forms. It may be a process, as is frequently the case in the metallurgical and chemical industries. It may be a machine or article of manufacture. It may even be a composition of matter—a blend or mixture that meets the *new and useful* criterion.

o   The   know-how   may   simply   be   an

improvement to an already patented process, machine, or composition of matter.

o   In general, mental processes or laws of nature are excluded.

o   Finally, some types of know-how are excluded by statute. For example, the government will not grant patents on certain defense-related know-how.

In addition, the law cites conditions that can cause a person to lose the right to a patent:

o   If the invention was known or used by others prior to the invention by the applicant, or in public use or on sale more than one year prior to the patent application in the U.S.

o   If the invention has been abandoned.

o   If there was a prior valid patent application on the same invention.

o   If the applicant himself or herself did not invent the subject of the patent.

Finally, the subject of the patent must possess *novelty*. In general, novelty is determined by whether the subject of the patent is sufficiently different from the prior art that it would not have been obvious to a person skilled in the art.

Note that the law requires full disclosure of the invention in the patent specification section. In order to ensure the availability of the invention to the general public at the expiration of the patent, the disclosure must be sufficiently clear to permit one skilled in the art to practice the know-how. The specification must conclude with specific claims that define the subject matter regarded by the applicant as his invention.

Copyrights

The field of technology transfer is concerned to a large extent with the above categories of intellectual property-- know-how, in general, and patents, in particular. However, because of their common characteristics of intangible value and legally sanctioned monopoly rights, it behooves us to take a very brief look at copyrights and trademarks.

A copyright can be obtained on literary and dramatic material, graphic and pictorial items, moving pictures, and sound recordings.   In general terms, copyrights apply to literary and artistic expression, just as patents apply to inventions.   As is the case with patents, certain exclusive rights accrue to the creator of the copyrighted work.   These include rights to reproduce and distribute the work, to carry out public performances, and to authorize others to carry out any of these activities.   If the material is published and distributed, it must carry a copyright notice, including the copyright symbol, year of first publication, and the name of the owner.

As in the case of patents, the exclusive rights accruing to the creator of the work have a finite life.   For material created during or after 1978, these rights expire 50 years after the author's death.   Items copyrighted before 1978 are protected for a period of 28 years, with a right to renew the term for an additional 47 years.

Trademarks

Most of us encounter trademarks on a daily basis in a multitude of commercial contacts.   We see them primarily in brand names, such as those of cigarettes, beer, clothing, and a host of other products, but also in other applications, such as certification, service, and collective marks.   Just as patents provide exclusive rights for invention and copyrights for expression, trademarks provide exclusive rights to marks of identification or commercial origin.

Actual rights to a trademark are obtained only by use and display.   However, for goods using the trademark in interstate or foreign commerce, the owner can register the trademark with the Patent and Trademark Office.   This registration, though not mandatory, can significantly enhance the exclusive rights associated with the trademark.   Those trademarks that are registered are usually identified with the familiar "Reg. U.S. Pat. and Tm. Off." or the familiar symbol ®.

The rights associated with trademarks will continue, in general, so long as the mark is used. Renewal of registration is required every 20 years throughout its use. If the mark is not used, it is considered to be abandoned after two years.

To summarize, technology transfer involves the effective management of valuable assets categorized as intellectual property. In many cases, these technological assets are unrecognized or, at best, undervalued, because they are intangible and sometimes difficult to characterize in quantitative terms. Proper recognition and management of these unique assets are essential for several reasons. First, through an effective licensing program, and contrary to other assets, the right to use technological assets is a product that can be sold repeatedly and, in some cases, concurrently, while still maintaining or enhancing the original value. Second, and again contrary to most other assets or articles of commerce, governments will provide monopoly rights for the commercial exploitation of these technological assets. There is clearly a strong incentive to develop an understanding of the techniques of effective technology transfer.

Part II

Technology Marketing and Out-Licensing

3

The Marketing Decision

The decision to market and transfer technology is, in reality, the result of a subset of related decisions. These can vary, depending on the product and the organization, but in general involve the following:

o   Organizational decisions

o   Product management decisions

o   Product selection decisions

Organizational decisions, for the most part, relate to our organization's attitude toward technology transfer as a business. *Product management decisions* address the issue of recognizing the existence of not one, but two products, i.e., a traditional, tangible product, such as a chemical or a resin, and a technology product, which is the know-how required to produce the tangible product. *Product selection decisions*, as the name implies, deal with picking the right products to package and market. Because product selection is so critical to commercial success, we shall place most of our emphasis here, with discussions on selection criteria and product categorization.

Let us now take a closer look at these decisions and the factors involved.

ORGANIZATIONAL DECISIONS

Let us consider an example. An organization has decided to capitalize on its technological assets which thus far have been used only in captive applications. A "minimum-effort" approach is selected, wherein one individual is named as the licensing manager. In order to avoid disrupting the existing organization, no direct assignments in R&D or Engineering are made. The licensing manager usually serves as a convenient referral for unsolicited technology inquiries that appear from time to time. In responding to these inquiries, he or she is forced to beg for support and assistance from Engineering and R&D organizations in preparing disclosures, process packages, and the like. Periodically, usually when in-house technical programs have slowed because of the general economy, someone will propose that "we should put more emphasis on licensing." This will generate a flurry of activity, some actual licensing proposals, perhaps even a license or two. Usually at this point the economy picks up, in-house technology support requirements increase, and the licensing activity becomes a costly embarrassment.

The point here is that a decision to market technology products must be accompanied by an organizational structure that can effectively support the technology transfer effort. Note that we are not advocating an "empire-building" effort. A large additional staff is not necessarily required. Rather, the critical issue is asset management and strategy definition. A decision to exploit technological assets must be a conscious one, with full-blown management support. Moreover, there must be a strong commitment of functional resources, such as R&D and Engineering, on a continuing basis. Finally, there must be a clear understanding of what the organization aspires to be in the world of technology transfer, i.e., source for a few selected products, or a full-service supplier within an industry.

PRODUCT MANAGEMENT DECISIONS

For an emotional response, there is little that can compare with a proposal to a product or general manager to license the technology for manufacturing his product to the world at large, perhaps *even his competition*. Such a suggestion, to put it mildly, is usually not well received. Yet, sooner or later, except in unusual circumstances, the issue must be addressed.

Let us start by accepting the premise that there is no one "right answer." There are some instances where marketing only one of the two products makes sense. For example, where technology is the only, or major, barrier to entry and cannot be obtained from other sources, a strong argument can be made for withholding the technological product.

At the other end of the spectrum of dual product management, there are instances where the proper decision is to market only the technological product. Bayer, for example, in West Germany, successfully developed an ethylene-based catalytic process for the manufacture of vinyl acetate. They had the choice of installing facilities and using the new technology to manufacture and market the vinyl acetate monomer itself, or simply marketing the technology as a licensed process. Bayer chose the latter course of action and successfully licensed the technology worldwide. Was this the right decision? Probably only Bayer can say for sure, but based on the financial yield from the licensing effort, compared to the capital cost and lack of a basic raw material position in vinyl acetate manufacture, the decision appears to have been a very good one.

The decision regarding marketing the tangible product or the technology does not have to be an "either/or" choice. To cite only one example, Monsanto, after developing its methanol carbonylation route to acetic acid, opted to market both the acetic acid and the technology. The acetic acid manufacturing facility at Texas City was an invaluable asset to the technology licensing effort. At the same time, Monsanto's position as licensor to a number of its competitors worldwide strengthened its competitive position in the acetyl product area. Effective management of these two related products can indeed provide synergistic results.

A decision to market both the tangible and the technological products must be accompanied by one other critical decision. This relates to the level of the organization at which there is common responsibility for the management of the two products. A rule of thumb, based on highly subjective reasoning, is that *the lower the level of common management responsibility for the two products, the more effective the overall marketing effort.* Obviously, the level must be high enough to ensure resource availability and functional support. However, one would hope to avoid a situation where a tangible product is marketed by one division, the technology for manufacturing that product by another division, and the lowest level of common product marketing responsibility is the chief executive officer of the corporation.

PRODUCT SELECTION DECISIONS

As in most businesses, the selection of the product(s) to be marketed is key to the success or failure of the enterprise. It is sad, but true, that an inferior product remains an inferior product, regardless of how it may be dressed up with a world-class promotion, advertising, and selling effort. Accordingly, product selection is one of the critical activities in the decision to market technological products.

If your organization is a sole proprietorship, or a small partnership or corporation, one individual may be able to identify and classify all available technological assets. Indeed, the product slate may consist of a single product, on which all the resources of the organization may be focused. As a matter of fact, the technological asset may be the organization's reason for being, and the understanding of its commercial value is likely the cornerstone of organizational strategy.

For larger organizations, however, the identification and selection of technological products for marketing can be a formidable task. No single individual is aware of all potential technological products, and seldom is there one convenient, comprehensive list. An obvious source is the list of patents owned or assigned to the organization, but concentration on this category to the exclusion of all else could be very shortsighted. Technologies that for one reason or another (prior commercial use, for example) may not be patentable may still have significant value to a potential licensee.

Usually, some sort of inventory effort can be beneficial. Circulation of a *brief* questionnaire, informal presentations to various management and staff groups, interviews with key individuals—all of these have been tried, with generally beneficial results. One will see a number of "favorite sons" and pet ideas, but there will probably also be some solid candidates that could otherwise go unnoticed and undiscovered. The technological areas (R&D, Engineering) are obvious sources, but do not overlook the commercial areas, such as the various marketing components. Their interface with the external environment can be invaluable in matching market needs with existing technological products.

From these sources and others that may come to mind, a number of possible technological products will be accumulated. The task then becomes one of classification. As a first-pass prelude to a more definitive classification effort, one might consider the following three-way split:

o   Stand-alone products

o   Add-on or supplemental products

o   Non-competitive products

Stand-alone products, as the name implies, would be those that can be marketed by themselves, such as a mechanical, metallurgical, or chemical process, or a particular product or composition of matter.

Add-on or supplemental products are those whose value lies in their contribution to the overall worth of a prime or stand-alone product. Examples would be catalyst regeneration or energy conservation techniques specific to a particular process. For the most part, the inclusion of these products in the marketing program will rise or fall with the disposition of the prime product.

We all know about non-competitive products. These are the technological products that lack novelty, utility, transferability, and intrinsic value. We all know about non-competitive products, that is, except for the product's sponsor or champion whose reaction can range from passive acceptance (really!) to anger and disbelief. Nevertheless, early identification of those products that lack commercial potential is essential to the long-term viability of the effort.

As a means to that end, let us consider some methods for screening the candidates.

Selection Criteria

In a later chapter, we will address the problem of placing a monetary value on a technological asset. Here we are concerned with more general techniques that can serve to evaluate the inventory of products discussed.

The following standards have been applied successfully in identifying technology products with the greatest marketing potential.

Value To The Licensee or Buyer. Imagine, if you will, a cold-eyed potential customer for the technology asking what this product will do for him. If you have trouble expressing in clear English (or preferably the customer's native tongue) exactly what beneficial results this technology will yield, there may be a significant message as to product value. Remember also that real advantages in your own operation may be non-existent or marginal for the customer, because of differing financial objectives or economic system characteristics.

In the final analysis, one must quantify the benefits to a potential buyer or licensee. An economic evaluation of the technology is highly desirable to define specific benefits, and also to ensure that marketing efforts are properly directed. Where there are known competitive products, an effort should be made to evaluate these for comparison purposes. Rest assured that the potential customer will undertake such a comparison.

Technology Transferability. Can you reduce to writing exactly what the technological product is, and how to produce it? Or does the practice of the technology require the continuing services of a veteran employee whose 30 or 40 years of experience in the practice of the technology have provided him with difficult-to-duplicate skills that are unlikely to be achieved by the licensee?

As will be seen in Chapter 5, technology transfer is accomplished by handing over prepared packages—descriptions, specifications, drawings, etc. These, of course, are supplemented by meetings and dialogue among technical specialists. However, years of experience and intuitive skills cannot be delivered as a neat package. If the bulk of what is required to practice the technology cannot be expressed in written or diagrammatic form, there is probably trouble ahead!

Resource Requirements. Let us assume your licensing efforts are successful. You have a genuine licensee, with a good possibility of a few more. The only problem is that you will have to divert most of your technical resources from internal needs to licensing support in order to meet contractual commitments. No crystal ball can define the number of licenses, but there should be a clear understanding of technical resource requirements for each technology transfer and, just as importantly, an assessment of the impact on internal technology needs.

In this regard, the problem of conflicting resource requirements is sometimes addressed by proposing what is usually labeled a "minimum-effort" technology transfer approach. This generally entails the preparation and delivery of the *minimum* contractually required technology packages and support to the licensee and/or his contractor. The result can be serious technical and commercial problems, together with a loss of credibility as a reliable technology supplier. Word of a badly-handled technology transfer gets wide circulation and can have long-term consequences. Remember that product quality standards are just as important for technology products as for others. If the technology transfer resources for a particular product are too great for the organization to supply, the product is probably not a viable candidate.

There are probably other standards and criteria that can be applied. However, the above three have been applied successfully to various product collections and can serve as an effective screening or selection process.

Product Categories

At this point, there should be a rather heterogeneous collection of stand-alone technology products that have satisfied these preliminary screening criteria. For purposes of resource allocation and the development of a marketing plan, some differentiation among these various products is desirable. A number of classification systems can be applied, but for preliminary planning purposes, the following categories have been found useful.

License As-Is. These are products that have satisfied the criteria of customer value, transferability, and reasonable resource requirements. These can be the subjects of current marketing efforts and can serve to establish a position in the target markets. In the best-of-all worlds, there would be a mix of small products with broad markets, plus a few larger technologies with less-frequent sales potential. The combination, of course, provides a load-leveling effect that is highly desirable. Unfortunately, this is a criterion that is sometimes approached, but seldom achieved.

Modify to License. These are existing technological products that can be made competitive and marketable with definable and achievable improvements by Engineering or R&D. These would be incremental process improvements to overcome specific competitive deficiencies, as opposed to massive technology overhauls. Items such as energy utilization, product recovery, and product quality would be typical considerations for improvement.

Sell. Certain technological products may fall into an "oddball" category because of applicability in a different marketing area, shortage of technical resources, or other deviations from the main product slate. If it is indeed a workable, competitive technology, and if it has no captive application, it may be a candidate for sale, as opposed to license. Salability is enhanced, of course, if there is a valid patent position.

Selling a technology, as contrasted to licensing the right-to-use, is particularly applicable if its principal use is in a completely different market from one's main product lines. The technology may be of considerably more value to a buyer already participating in that market. Even though the owner may have some current or future use for the technology, a sales contract that provides a royalty-free license to the former owner can still provide needed access while permitting maximum value to be realized for the technology.

Acquired Products

Thus far, we have assumed a product-driven product selection process based on technological products available in-house. As indicated previously, and also in subsequent chapters, these products must also pass muster regarding market needs, competitiveness, and transferability. However, the initial selection has been based on products available within the organization.

Other sources of technological products are, of course, outside individuals, organizations, and agencies. More detailed information and background on the acquisition of technological products from outside sources, either for use or licensing, is presented in Part III: Technology Acquisition, and we will not discuss approaches and methods at this point. Nevertheless, some thoughts on the use of purchased technology for a licensing program may be appropriate. Unless the technology from outside sources is a good fit with, or supplement to, the in-house technological product slate, tread carefully. Keep in mind that these "strangers" entail another technology transfer step, with all the risks and resources related thereto. In addition, there are sometimes credibility questions related to marketing technology developed by others, particularly in competitive selling situations. Acquisition of licensable technology from other sources can serve to strengthen and enhance the value of in-house technology in special situations. These are usually instances where there is a gap or a weakness in an in-house technological product that can be filled or remedied by the purchased technology.

In summary, the decision to market technological products is a complex one. Questions, such as directed versus opportunistic marketing, product management structure, product identification and classification, and organizational requirements, must be addressed and resolved. Above all, there must be a common understanding of what the organization aspires to be in the field of technology transfer, and how that role dovetails with the overall strategy of the organization.

4

Market Research

Market research usually brings to mind extensive demographic evaluation, regression analyses of past market behavior, and comprehensive sampling and interviewing techniques. Many of these techniques, however, are oriented toward consumer behavior and response, low-priced commodities, and easily transferable, tangible products.

With technological products, the ground rules and market research needs are very different. For one thing, contrary to many consumer tangible products, it is more difficult to *create* a demand for technological products. The demand arises from other forces, such as market demand for tangible products produced by the technology, state planning or incentives, or perceived commercial opportunities unrelated to the technology. It is probably obvious, but sometimes overlooked, that technological products are truly means to an end, rather than the desired end itself. One can influence to some extent, the demand for soft drinks or fashion products; it is unlikely that one can influence or accelerate the perceived need for an ethylene plant unless other factors, real or imaginary, have signalled the need.

There are other differences affecting the kind and extent of market research required for technological products. Relative to consumer products, the decision-makers for purchases of technological products are few. Prices are orders of magnitude higher, and consequences are long-term. One might argue that purchase decisions on technological products are made on a more rational, objective basis; most of us, however, can cite a number of "gut-feel" purchase decisions on licensing issues. Certainly, the decision-making process is longer and more complex; we will leave to others the determination of whether the additional deliberation and analysis lead to superior choices.

Because of these considerations, market research as used in this book in the area of technology transfer is different from that for tangible consumer or industrial products. Perhaps the best way to define the activity is in terms of the questions it should answer, as follows:

> o   Who are the likely buyers (or licensees) for our technological products? How do we best screen the potential buyers so as to focus our marketing efforts?

> o   Once these likely buyers are identified, how do we most effectively market our technological product to them?

> o   What is the competition, and how does our product compare to the opposition?

Let us take a closer look at these questions, as well as how to go about developing answers that are useful in structuring a marketing plan.

LIKELY BUYERS AND LICENSEES

One approach to the identification of these potential buyers would be to screen carefully the journals and other media outlets for announcements of new facilities that could use our technology. In most cases, however, this approach could be a classic case of "too little and too late." By the time an announcement is made, technology selection has been completed, or is already well under way.

It is usually far more effective to develop a list of possible users of our technology. Some screening criteria that can be used to develop such a list are as follows:

o   Does the potential buyer already have a
     position in the market areas served by
     tangible products of our technology? A
     company already participating in the
     industrial solvents market is a far more
     likely candidate for ketones or esters
     technology than one whose principal
     market focus has been fertilizers and
     related agricultural formulations.

o   What is the raw material position of the
     prospect? Who is more likely to license
     ethylene   derivative   technology—a
     manufacturer with existing underutilized
     ethylene   production   capacity,   or   a
     manufacturer with no existing ethylene
     supply whatsoever?

o   Is the prospect already manufacturing
     the product of our technology, albeit by
     a   different   technology?   If   our
     technology is truly competitive, this is
     indeed a prime prospect. The company
     may not be interested in discarding its
     captive technology, but will consider
     competing technologies for the next
     increment of production.

By using these preliminary screening criteria, one can
put together a list of likely prospects, at least for the
domestic U.S. market. Keep in mind, however, that there
will always be exceptions to the rule, i.e., prospects who lack
market and/or raw material position, but have other strong
motivation to enter businesses that could require our
technology.

Outside the U.S., where many technological products
may find a broader market, the development of a likely
prospect list is more complex. Certainly the same strategic
issues apply: market position, raw material position, and
general business outlook. Overriding all of these
considerations, however, are the methods by which the
nation's economy and its growth and direction are planned
and implemented.

Because of this, an understanding of some of the

economic characteristics of a target country is desirable. Becoming an expert or scholar on the details of a country's economic system is not necessary, but it is essential to realize that the areas of interest are the level of technological sophistication, the role of government in industrial growth, and the production facilities and markets in the industrial sector(s). Of particular interest is the planning mechanism, as well as the access to economic plans and programs, particularly those based on the well-known five-year planning cycle.

Be mindful of the fact that one cannot simply divide the nations of the world into two tidy categories—free enterprise and state-planned. Think of any gradation between these two categories, and it will probably match some country's economic system. Although both Hungary and Bulgaria, for example, are Eastern Bloc economic systems, they are far from being clones in their approach to industrial planning and development.

There are a number of other sources that can be of value in identifying countries that would be likely prospects for a particular technological product. Obviously, if your organization has facilities or offices in a particular country or region, these represent one of the more direct and effective approaches. In many instances, these representatives will have developed personal contacts with decision-makers and, just as importantly, understand the government-industry interface and the planning and direction-setting mechanisms.

Another market research resource for international technology transfer that is sometimes overlooked or underutilized is that of U.S. government agencies, particularly the Department of Commerce. The specialists in the department's Washington offices who manage the various country "desks" can be an invaluable asset in gaining an understanding of a country's industrial and commercial structure. In addition, the Commerce Department has International Trade Administration Offices in many cities which are "wired-in" to the department's specialists and other resources. For example, if one is interested in marketing technology related to the vinyl resins market, the International Trade Offices can access data banks to identify industrial organizations in this market in particular countries.

For each of these companies, information is provided regarding key management personnel, sales volumes, number of employees, and facility location. Another excellent non-government source of information for its members is the U.S. Chamber of Commerce and its staff of country and regional specialists in its Washington Office.

Finally, do not overlook available published resources. Although unlikely to include "tailor-made" data on your area of interest, a good general overview of a country or region can be obtained from most libraries' general reference section. It may seem somewhat academic, but useful overviews can be obtained from such standard reference works. Particularly useful reference sources, where available, are the subscription studies carried out by consulting firms, such as Stanford Research Institute, Chem Systems, and others, on the worldwide outlook for particular technologies in various product areas.

In the screening process, particularly in international markets, be alert for market niches that, because of local circumstances and national goals, might be well served by a particular technology. Consider, for example, a Third World country with an agricultural economy, soft currency, and aspirations toward a "minicomplex" of various ethylene derivatives, such as polyethylene resins, ethylene glycol for textile manufacture, and others. Balance of trade and diseconomies of scale would preclude a conventional hydrocarbon-fed ethylene plant to supply the derivatives units. A considerably more attractive prospect would be a small fermentation unit converting agricultural products to ethanol, followed by conversion of the ethanol to ethylene for use in derivatives. Indigenous raw materials, low capital requirements, and reduced penalties for small scale operation—these are effective selling points for the marketer or licensor of these technologies in what is admittedly a relatively small niche in the worldwide market for ethylene and derivatives technology.

Finally, a few words regarding "state of industrial development" as a measure of market potential in a country for a specific technology product. State of development can be an effective screening device when accompanied by a realistic assessment of the technical and facility support requirements for the practice of the technology. Are we looking for highly developed industrial nations, or the less developed countries (LDCs)?   Depending on our particular technology products, the answer may be neither, but some intermediate state of development. The ideal market may be a country with a need for using domestic raw materials and minimizing foreign exchange requirements, yet with sufficient industrial structure (electric power, transportation) to support an industrial facility based on our technology product. Application of this one criterion could narrow the field considerably.

With the application of the criteria outlined in the preceding pages, a "hit list" of corporations and countries that could be potential licensees for our technology can be developed.   Once we have such a short list, we are faced with the second of the questions posed at the beginning of this chapter.

EFFECTIVE APPROACHES TO MARKETING

In the best of all possible worlds, a group of experienced marketers, backed by excellent technical resource people, would be assembled to work on the selected targets. Although there have been a few instances of such a grandiose approach, a far more likely response from upper management sources has been the familiar "make do with what you have!"

In the spirit of making do with what we have, or perhaps just a bit more, let us consider some workable approaches.   Because of the additional complexity, the principal emphasis will be on serving international markets; however, the principles and approaches are generally applicable also to domestic markets.

First of all, make use of existing commercial contacts. They know the territory, they know the organization and its people, and they know (in some cases) something of the potential customer's future plans and objectives.   Existing commercial contacts, of course, can mean different things to different people.   In general, we are referring to the management of local sales offices, account managers, marketing managers—the job titles will differ, but the common characteristic is a previously established commercial interface with the potential customer for our technology.

In addition to the commercial background and experience gained by the use of existing commercial contacts, there is another significant benefit. By involving this organization, there is a greater likelihood of an integrated approach to the marketing of tangible products (electronic components, chemicals, etc.) and technological products (know-how, license to practice, etc.).

It should be emphasized that the involvement of existing commercial contacts does not mean dumping the technology marketing and sales effort into the existing marketing organization. Technology is a vastly different product, requiring different kinds of marketing support and entailing different points of contact with the customer. Even so, the existing marketing organization, when available, can be an invaluable asset in establishing initial contacts and credibility.

Another effective approach to reaching or identifying potential customers for technology is the use of third parties, notably engineering contractors. In many cases, a joint effort in technology transfer between the owner of the technology and an engineering contractor can be a highly effective technology marketing team. The owner has the technology, technical personnel skilled in the technology, an operating facility, and a need for technology marketing muscle. The engineering contractor has an experienced marketing staff, in many cases a presence in the market areas of interest, the capability of building facilities in which to produce the technology, and a need for technological bases for his facility proposals. Like a good marriage, the joint efforts in technology transfer can be a "win-win" situation. Unfortunately, they can also be a series of pitfalls for the unwary, with significant penalties for both sides unless the effort is carefully managed from the outset.

Some of the areas to be considered in structuring such a joint effort in technology transfer follow:

Know Thy Contractor

A clear understanding of the engineering contractor's capability and, in particular, his track record in marketing engineering service in the areas of interest is a must. Potential customers' or licensees' perception of the contractor is critical. In addition, an understanding of the contractor's marketing organization, resources, and regional offices is highly desirable. A very significant consideration, but one often overlooked in evaluation, is the contractor's access to financing for new facilities and technology in various geographical areas.

Define the Areas of Responsibility

This sounds obvious, and admittedly very basic, but the areas of responsibility must be resolved at the outset. It is very painful (and sometimes costly!) to deal with these issues in the midst of commercial discussions with a potential licensee. Technical review responsibility, for example, can be a critical issue. There is sometimes a tendency for a licensor or technology owner to opt for a minimum-effort participation with a contractor and to shirk technical involvement and review. A classic case in the 1970s was a minimum-involvement effort by a U.S. licensor in reviewing a contractor's engineering efforts on an Eastern European facility. The fix-ups and remedial action required from the licensor following a disastrous start-up and initial operating period far exceeded the commitment that would have been required for a first-class monitoring and review effort during the contractor's detailed design efforts.

In addition to the technical review effort, there should also be a clear understanding of other technical support requirements, guarantee commitments and penalties, start-up and commissioning support, and marketing participation.

Do Not Abdicate Commercial Involvement

There may sometimes be a sense of relief that the engineering contractor will handle all of the grimy details of negotiating a technology transfer and facilities agreement with the customer. However, unless there is involvement by the technology owner or very specific guidelines spelled out in advance, the differing objectives of the technology owner and the engineering contractor can yield some king-sized problems. A particular problem area is that of lump-sum proposals for the combined facility and technology license. When a price concession is negotiated, how is this allocated between the facilities price and the license fee? Who can adjust guarantee levels during negotiations, and how are guarantee penalties distributed? All these can be resolved, but much more easily in advance of the pressures of negotiating sessions with the potential customer.

Except for Special Cases, Avoid Exclusivity

There are a lot of competent international contractors, each with some access and particular strength in a difficult market area. In addition, a cherished prerogative of many potential customers is that of selecting their own contractor. To make the use of one particular contractor a prerequisite for using your technology is to sharply restrict your marketing effort.

In some cases, there have been joint technology marketing efforts between a technology owner and an engineering contractor, with at least implied exclusivity. These have generally evolved when there have been technical contributions on the part of the engineering contractor or a joint technology development effort.

Some technology owners have found it beneficial to establish a non-exclusive, but "most-favored-contractor" relationship with selected engineering contractors. In this arrangement, the licensor or technology owner provides the contractor with disclosures on his technology and sufficient technical input to familiarize the contractor's technical and marketing personnel with the product. The selected contractors then have the advantage of being able to respond to inquiries more expeditiously and effectively without a time-consuming technology transfer process from owner to contractor. Predetermined procedures and communications requirements apply to the contractor's response, so as to ensure involvement and commitment of the technology owner.

As noted earlier, the third question that must be addressed in our concept of market research is the familiar refrain:

WHAT IS OUR COMPETITION, AND HOW DO WE COMPARE WITH IT?

In general, the first response to the question is that our competition is any other technology that can permit the user to produce the same (or a similar) product. Care must be exercised, however, to understand the true objectives of the inquiry when defining the competition. The competition may well be continued imports of the product, or it may be technology for a completely different product that will produce similar amounts of needed foreign exchange. The effectiveness of the response to an inquiry will be proportional to our understanding of the true competition and the real objective of the organization making the inquiry.

Within the narrowest context of competition, that of various technologies to produce the same or similar products, it is necessary to know more than the identities of competitors. Rather, there must be an understanding also of the financial competitiveness of our technology in relation to the competition. Analyzing the economics of competitive processes is by no means an activity unique to technology transfer. In fact, in many organizations, it is an evergreen activity and an essential element of the strategic business planning process. However, the competitive analyses required for technology transfer may differ in some respects from the conventional return on investment or discounted cash flow evaluation.

First, there may be heavy emphasis on capital cost, as opposed to annual operating or manufacturing costs. This is particularly true where foreign exchange is a problem or where financing is a critical element. Traditional trade-offs between capital costs and operating costs will probably not apply, and a competitor's offer featuring low capital cost with what would appear to be prohibitive operating costs may be received with the equivalent of a standing ovation.

Second, operating costs that appear to be unacceptably high relative to world-scale product prices may not be as big a problem as in more industrialized nations. This is particularly true where the proposed facility will be the first of its kind in the area, and the borders will be closed to imports of that product subsequent to facility start-up.

Third, and perhaps obviously, labor savings and automation will not weight heavily in the quantitative evaluation of a technology where local labor rates are low and unemployment is high. Tread cautiously here, though, because the implication of an offer of obsolete labor-intensive technology can strike a nerve with very negative results.

All in all, an effective and useful analysis of the competition will require knowledge and understanding of something more than the basic technology and conventional cost factors. The competition analysis must be tailored to fit local conditions, perspectives, and policy issues.

In summary, there are ways of defining and screening potential customers for a technology product, so as to focus the marketing effort. Both internal and external resources can be applied in reaching these selected markets, but care must be exercised to maintain sufficient control to ensure effectiveness. Finally, an active understanding and careful economic analysis of the competition can be an essential marketing tool in technology transfer.

5

Product Packaging

Somehow, *packaging* evokes thoughts of wrappers, cartons, cans, and other forms of potential litter. Probably necessary, but not always the key element in product delivery, packaging sometimes receives very little consideration, particularly in the hard-goods area. With tangible goods, packaging is important as a marketing device and as a delivery container, but is technically distinct from the product itself.

In technology transfer, the role of packaging is very different. In past chapters, we have seen that technology products, such as know-how, whether patented or not, are intangible species of intellectual property. For these products, it is necessary to develop some physical representation in order to market and deliver the product. In a sense, the package becomes the product, at least in the eyes of the beholder, who may be a marketing prospect or an actual licensee of the product. In view of this, product packaging becomes a very critical element in the whole process of technology transfer.

In the whole area of product packaging for technology transfer, one issue arises repeatedly. This is the continuing dilemma of providing sufficient information for product evaluation by the customer, without thereby also giving the product away. If we are selling shoes, or athletic equipment, a potential customer can try them on, or try it out, to determine fit and suitability. Even with consumables, a descriptive label and the customer's past experience can serve to characterize the product sufficiently for a purchasing decision.

Contrast this product characterization for hard goods with that for a technology product, such as process know-how. We can tell our potential customer all the truly great features of our product and let him see it in action in an operating facility. However, if we are not very careful, we may find at the end of the effort that we have inadvertently delivered the product with no quid pro quo, such as a license fee or royalty payments. On the other hand, a policy of tight secrecy regarding the nature of our product is going to be a very tough sell. Most potential customers do not reach for their checkbooks unless and until they have a pretty clear picture of what the product is going to do for them.

Because of this dilemma, packaging of technological products has become a very critical element in the success or failure of the total technology transfer program. Although there is no absolute standard as to what constitutes a suitable package for a particular purpose, most technology transfer efforts entail two major functions: *marketing* and *transfer*. Accordingly, current practice in technology transfer is to classify technology packages into these two major functions. These classifications are summarized in the table below:

Technology Products Packaging

A.   Marketing Packages

1.   The "black box"

2.   The non-confidential disclosure

3.   The confidential disclosure

B.   Transfer Packages

1.   The process package

2.   The detailed design

3.   The production facility

As the name implies, marketing packages are tangible representations of the technology concept, designed primarily to provide a potential customer with sufficient information to determine the worth of the technology. These are partial disclosures, with varying levels of detail and differing secrecy constraints, all used before a potential customer has executed a license agreement. In general, a marketing package does not contain sufficient information to permit the design and operation of a facility in which to practice the technology.

The transfer packages, on the other hand, are designed specifically, as the name implies, to transfer the technology to a licensee of the technology. The different levels of package completion are representative of those applicable in the chemical, refining, or metallurgical fields. As noted in the tabulation above, the ultimate technology transfer package would be the turnkey commercial production facility itself.

Let us now take a more detailed look at these various forms of technology packages.

MARKETING PACKAGES

The "Black Box"

So called because it conceals the process details as though hidden in a black or darkened container, the "black box" is the simplest form of technology package. Nevertheless, it does contain sufficient information so that one can make a preliminary evaluation of the economic worth of the technology. Black box disclosures are usually not tailored to a specific inquiry and, in many cases, are off-the-shelf items that are available on very short notice. In addition to satisfying specific inquires, black box disclosure packages are often used as the basis for publicizing the technology in print media, such as trade journals and advertising brochures.

In content, a black box disclosure will occupy one or two pages at most. For most packages, it will vary depending on the technology and its complexity, but usually a block flow diagram with a block for each major system is included, showing major raw material, intermediate, and product flows for a stated capacity. In addition, the black box disclosure package will display requirements for operating labor, utility and energy components, miscellaneous raw materials, and maintenance. An order-of-magnitude estimate of the capital cost of constructing the facility will also be presented, referenced to a particular production capacity, geographical location, and construction period or inflation index. In some instances, a paragraph or two discussing the process and its advantages and unique features are also included.

An illustration of a typical format for a black box disclosure is given below.

E X A M P L E

(BLACK BOX DISCLOSURE)

EUREKA PROCESS

Production Capacity   100 Million lb./yr.

Operating Requirements   Per Pound of Product

    Raw Material

    Utilities

       Steam   _____

       Cooling Water   _____

       Electricity   _____

    Catalyst   _____

Operating Labor   ___ positions/shift

Maintenance   ___ percent of fixed investment

Capital Cost (Battery Limits)   US $_____

Basis: US Gulf Coast

1987 Construction Costs

The Non-Confidential Disclosure

This technology package is more comprehensive than the black box disclosure but, as the name implies, it does not convey confidential information. It may, to some extent, be developed or modified to fit a specific request from a potential licensee. Although brief, it will be used in many cases to provide background and perspective on the technology owner's capability and technical expertise, both in general, and also with regard to the specific technology. Length of the disclosure package will vary, of course, depending on the complexity of the technology and the level of the marketing effort. Format of the package varies from a stapled collection of plain bond paper to a bound, indexed, and personalized pamphlet with multicolored printed covers. Most experienced licensors recognize the role of the non-confidential disclosure as an effective marketing tool, and develop the packages accordingly.

The content of the non-confidential disclosure is affected greatly by the amount of information that can be disclosed without compromising confidentiality. Compared to the black box disclosure, the non-confidential disclosure will contain significantly more information on the licensor's capability in transferring the technology and on the process features and characteristics. In many instances, the flow diagram will show major items of equipment in outline form, but without sizes, materials, operating conditions, or flow rates. Sufficient information will be included to permit the reader to determine raw material, steam, water, electricity, and miscellaneous material requirements. An estimate of capital cost for the desired capacity will be included, along with the basis for the estimate and a definition of what is included. In some cases, man-hours of construction labor, indexed to a given productivity level, will be broken down to permit adjustments for a particular economy. Adjustment factors for different design capacities are also included sometimes, depending on the nature of the inquiry.

Summarized on the following page is an outline of a typical non-confidential disclosure package. Note that significant changes in format and content may be made to accommodate specific inquiries or technologies.

Note that there is no information on commercial terms

OUTLINE

(NON-CONFIDENTIAL DISCLOSURE)

I.   Introduction (Overview of licensor experience)
     A. Specific experience in the industry
     B. Experience in licensing this technology and others
     C. Technical resources
     D. Support services
II. Technology-Specific Background Information
     A. Licensor experience
     B. Process description
     C. Past technology developments
     D. Operating facilities
     E. Other licensees
     F. Unique or advantageous features
III. Evaluation Data and Information
     A. Raw material(s) specification
     B. Product quality and specifications
     C. Operating requirements (per unit of production)
         1. raw material
         2. steam
         3. cooling water
         4. electricity
         5. miscellaneous material
     D. Other requirements
         1. maintenance
         2. operating labor/staffing
         3. capital cost (battery limits, referenced to a
             geographical area and construction period)
IV. Flow Diagram
     (Major equipment only, no operating conditions, process
     conditions or equipment sizes)

included in the technology package. This is obviously not a
hard and fast rule and can be altered to fit circumstances.
However, it is usually advantageous to supply terms only on
request and based on as much information as possible about
the potential licensee's economic situation and level of
interest. Certainly, if the inquiry includes a request for
commercial terms, an outline of these can be included in the

cover letter transmitting the non-confidential technology package.

The Confidential Disclosure

As the name implies, this disclosure places secrecy constraints on the recipient, usually in the form of a secrecy agreement executed prior to disclosure. The amount and type of information disclosed in a confidential disclosure, over and above that disclosed in a non-confidential disclosure, is by no means standardized. Rather, it is determined by what are considered to be the key elements of the technology that must be diligently protected to prevent unauthorized use.

In general, a confidential disclosure does not contain sufficient information to permit the recipient to practice the technology. The purpose of a confidential disclosure is to provide enough information to permit a more comprehensive evaluation of the technology by the recipient than could be done with a non-confidential disclosure. In order to do this, certain aspects of the technology, depending on the product, must be disclosed for the evaluation. It should be remembered, however, that the confidential disclosure is still a marketing package and not a transfer package.

On the following page is an example outline for a confidential disclosure. Again, it should be noted that the difference between a non-confidential disclosure and a confidential disclosure is highly dependent on the nature of the technology and what are considered to be key proprietary components of the product. In this example, the principal differences are:

o   A more detailed description of the process, highlighting unique features and significant advantages.

o   More emphasis on safety and environmental aspects of the technology.

o   A discussion and breakdown of the capital cost estimate.

o   A rudimentary major equipment list that can permit the recipient to do his own capital cost estimate, reflecting his own site conditions and also potential equipment import and foreign-exchange issues.

OUTLINE

(CONFIDENTIAL DISCLOSURE)

I.   Secrecy Constraints

    A.   Summary of obligations under previously executed secrecy agreement

    B.   Reference to secrecy agreement usually noted on each page of disclosure

II. Introduction

    Overview of licensor experience, similar to that used for non-confidential disclosure

III. Technology-Specific Background Information

    Information regarding the particular technological product. More detail on the technology itself in order to emphasize and validate points of superiority

IV. Evaluation Data and Information

    Information included in non-confidential disclosure, plus:

    A.   Discussion of staffing philosophy with sufficient detail to permit reader to adjust for local conditions and policies

    B.   More information on operating requirements (raw material, energy, labor) for sub-systems of facility

    C.   Detail on major elements of capital cost estimates

It is, perhaps, obvious that the recipient of a confidential disclosure who has chosen to obligate himself to secrecy is indeed a very serious and interested potential customer for the technology. For this reason, the form and appearance of the disclosure is quite important. Although disclosure appearance will not overcome flaws and omissions in the content, it nevertheless conveys a strong message to the recipient. This message makes clear the attitude of the licensor regarding technology transfer in general, and implies the level and quality of technical support that one could expect. Accordingly, the effort and time required to create a professional appearance for the disclosure yields significant dividends in marketing results.

One common question with regard to confidential disclosures is whether to charge a fee for the disclosure. As in other areas of technology transfer, there is no one correct answer for all cases. One argument is that the confidential disclosure is indeed a marketing package, and the imposition of a fee for the preparation and delivery of such a package impedes the marketing effort by discouraging potential licensees. The counter-argument states that the imposition of a nominal fee, such as $100,000 or less, serves to ensure that those who do take the disclosure are serious licensee candidates. In general, the imposition of a disclosure fee appears to be a function of the competitive situation. Where the technology is new, patent-protected, and one-of-a-kind, disclosure fees are more common than where the technology is merely one of several that can be used to produce the same product or result.

TRANSFER PACKAGES

Early in this chapter, it was noted that technology packages, representing tangible manifestations of intellectual property, performed two primary functions: technology marketing and technology transfer. We have discussed the three principal types of marketing packages, along with their content and use. We now turn to a consideration of the other function, that of actually transferring the technology.

In the introduction to the area of product packaging, we listed three versions of transfer packages:

o   The process package

o   The detailed design

o   The production facility

From the licensee's standpoint, the technology can be delivered in any of these forms. He may receive a process package containing all necessary technical information and proceed to implement the design and construction of manufacturing facilities himself. Alternatively, he may choose to work with the licensor and/or an engineering contractor and to receive the technology in the form of detailed design drawings and specifications, representing the input of the various technical disciplines in translating the technology to physical terms. Finally, he may opt for a turnkey manufacturing facility, wherein the technology is delivered in the form of ready-to-operate hardware. However, in each of these instances, the key technology transfer device is the process package, whether it is transferred to the licensee or to an engineering contractor for detailed design and construction. Accordingly, we shall confine our discussion of transfer packages to the process package.

The Process Package

In the chemical and refining industries and, to a large extent, in the metallurgical area, the key technology transfer instrument is the process package. Reflecting, as it does, a substantial portion of the know-how required to practice the technology, the process package is developed and delivered after a license agreement is executed.

As in the case of the other technology packages discussed heretofore, the content of the process will vary, depending on the nature of the technology and the needs of the licensee. A typical outline of a basic process package is presented on the following pages.

OUTLINE

BASIC PROCESS PACKAGE

I. Design Basis

    A. Capacity

    B. On-stream time

    C. Product mix

OUTLINE

BASIC PROCESS PACKAGE

(CONTINUED)

    D.   Raw material specifications

    E.   Product specifications

    F.   Efficiency

    G.   Chemicals and solvents required

    H.   Raw material usages

    I.   Utility supply conditions

II.   Basic Design Data

    A.   Site conditions

    B.   Raw material conditions at battery limits

    C.   Utility conditions at battery limits

    D.   Product conditions at battery limits

III.   Process Description

IV.   Process Flow And Control Diagrams Showing:

    A.   Major equipment

    B.   Stream data

        1.   flow rates

        2.   compositions

        3.   temperatures

        4.   pressures

        5.   heat duties

        6.   key stream physical properties

        7.   other critical data

    C.   Critical instrument loops

    D.   Critical dimensions and line sizes

V.   Major Equipment Specifications

    A.   Similar to that used for purchasing

<u>OUTLINE</u>

BASIC PROCESS PACKAGE

(CONTINUED)

    1. critical vessels - mechanical design drawings

    2. heat exchangers, columns - functional
       specifications

    3. performance specifications

       a. pumps

       b. compressors

       c. trays

       d. standard vessels

  B. Description and guidelines for materials of
     construction

  VI. Minor Equipment Specifications

 VII. Plot Plan Showing Major Equipment and Critical Dimensions

VIII. Operating Considerations

  A. Normal

  B. Start-up

  C. Shutdown

  IX. Utilities Requirements

   X. Waste Streams

  A. Quantity

  B. Composition

  C. Pressure

  D. Temperature

  E. Physical Properties

  XI. Special Considerations

  A. Special insulation specifications

  B. Special piping specifications

OUTLINE

BASIC PROCESS PACKAGE

(CONTINUED)

    C.   Special layout considerations

    D.   Special cleaning requirements

XII.   Safety Considerations such as:

    A.   Specific piping arrangements

    B.   Material residence times

    C.   Emergency venting requirements

    D.   Potential process hazards

    E.   Potential operational failures

    F.   Electrical area classification criteria

XIII.   Equipment List

XIV.   Instrumentation, including

    A.   Critical process control concepts

        1.   interlocks

        2.   bypasses

        3.   alarms

    B.   Logic diagrams for emergency shutdown

    C.   Instruments requiring emergency power

SUPPLEMENTAL INFORMATION

Physical Property Manual

Contains physical properties for all major components, plus critical mixtures

OUTLINE

BASIC PROCESS PACKAGE

(CONTINUED)

Standard Practices Manual

Outlines maintenance and operating practices and

procedures

Analytical Manual

Contains specifications and analytical methods for the

technology

This basic process package is usually converted by the licensee or his contractor to a more detailed package suitable for hardware design and procurement. The essential technology transfer, however, is implemented by the basic process package.

Because it is the key element in the transfer of the technology, the process package is covered by a secrecy agreement. If the process package is being delivered directly to the licensee, the confidentiality clauses in the previously executed license agreement itself may be sufficient. However, as is often the case, additional secrecy agreements will be required if a third party, such as an engineering contractor, is involved.

It should be noted that these transfer packages do not usually constitute the total technology transfer effort. They must be accompanied by access to, and support by, knowledgeable technical personnel in the licensor's organization. No document, complete and well-written though it may be, can substitute for input and response from experienced engineers and technicians. These supporting services and input will be discussed further in Chapter 10—Licensing Support Activities.

Process packages are usually priced separately from other technology costs, such as license fees and running royalties. This is done, in part at least, to provide flexibility in tailoring the process package to fit the licensee's needs. Because the cost of package preparation will vary, depending on the number of times the package has been prepared, costs are usually quoted on a lump-sum, rather than actual cost and material, basis so that each licensee will pay an equivalent amount. Otherwise, the first licensor, who may be the key customer, would bear the brunt of the cost.

The pricing philosophy is usually directed toward recovering fully absorbed (with overhead) costs for the first-time preparation of the package for each licensee. Exorbitant package costs with significant profit margins are usually avoided, because the process package is in reality a means of achieving a desired result, i.e., an effective technology transfer. Profits from the venture will be realized on the license fees and royalties, but only if the technology transfer has been handled in an effective and professional manner.

To review, product packaging is an extraordinarily critical element in technology transfer. Intellectual property, such as technology and know-how, must be packaged effectively to facilitate any kind of marketing effort. Examples of these marketing packages are the black box disclosure, the non-confidential disclosure, and the confidential disclosure.

Packaging also plays a vital role in transferring the technology. The key transfer package is the basic process package, containing sufficient information for the licensee or his contractor to design and construct a facility in which to practice the technology. Other forms in which the licensee may receive the technology are a detailed engineering design package or a ready-to-operate turnkey version of the facility itself. All, however, are based on the fundamental technology transfer device, that of the basic process package.

6

Promotion and Advertising

"Are you serious? Promotion and advertising in technology transfer? Promotion and advertising are for soft drinks, cigarettes, and show business! If your technology is good, it'll sell itself. If it isn't, nothing will help! Let's just concentrate on improving the technology."

This is not an unusual or overstated position. As a matter of fact, it is a paraphrased version of a statement made during the preliminary planning and development efforts on a new licensing program. It is a fair reflection, in fact, of what was described in Chapter 1 as "The Better Mousetrap Syndrome," contending that all one needs is a good product, and the world will beat a path to the owner's door, clamoring to buy the product. Would that it were so! Sales and marketing resources could then be reallocated to order-taking and follow-up, thereby saving significant amounts of both human and financial resources. Unfortunately, competitive reality has shown time and time again that is simply will not work. Even with an outstanding product, there is a serious need to get the message across to a prospective buyer.

It is true, of course, that different approaches are needed for a technology product than, say, for a new brand of beer. Even the most creative television advertisement will not deliver the message or reach the audience needed to license a new metallurgical or chemical process. As we noted before, technology is a product for which it is very difficult to create a demand. Nevertheless, where the demand exists, for whatever the reason, a carefully planned, highly focused promotion and advertising effort can greatly enhance the odds for success.

We should also be clear on the meaning of promotion and advertising in the overall context of technology transfer. Promotion and advertising can involve the traditional activities, such as the use of print media or direct mail; however, promotion can also involve the not-so-obvious activities, such as exploratory calls on potential customers and articles in technical journals.

Regardless of the form of the effort, the key element of the undertaking is a clear understanding of the message to be delivered on each occasion. Regardless of past practices, it does not usually excite the client to learn the intimate details of your organizational structure or to get more than an overview of last year's financial performance. Even a picture of your founder does not usually bring forth much emotion. Ask yourself what the "take-home" message is and to whom it is directed. Are we trying to say the following?

o   We are a reliable source for a broad array of technology products.

o   We are well-equipped to provide technical support for any technology you license from us.

o   This technology has a significant advantage in economic terms over anything the competition may have to offer.

o   Our running royalty may appear to be high, but consider what you are getting, relative to the competition.

The definition of the desired message may appear to be rather obvious, but how often have you seen a promotional or sales effort, printed or personal, wherein the only apparent message is that the seller would very much like to sell you something?   In yet other instances, the commercial message may be nearly obscured by generalities extolling the organization's success in commendable, but unrelated, areas.

There are a large number of advertising and promotion approaches, but technology transfer efforts usually fall into three principal categories;

o   Print Media

o   Trade and Technical Events

o   Personal Contacts

These categories will be discussed in the following sections.

PRINT MEDIA

Let us start by recognizing the obvious: the likelihood of someone taking a license for a particular technology solely because of an advertisement or news item in a newspaper or journal is almost non-existent.   Other forms of printed material, such as brochures or mailings, would fall in the same category.   Nevertheless, these can be useful contributors to the overall effort of creating an image as a competent marketer of technology products.

Print media efforts can vary from widely distributed press releases pinpointing advances in a particular technological product, to a broad-brush institutional overview of your organization's technology orientation.   The specific role that print media will play in the overall technology transfer marketing effort, of course, will be determined by the business strategy and focus for the effort.   If our marketing effort is directed toward a relatively few potential customers, with a very limited product slate, the broad coverage achievable through the use of conventional print media is probably not warranted.   Nevertheless, print media in its various forms does offer an effective approach to broad coverage of selected audiences.   Consider some of the approaches.

Press Releases

Sometimes we tend to overlook several significant facts

regarding the use of press releases as a means of getting a technology transfer message across to a selected audience. First, many of the activities related to your efforts in the technology transfer area are of considerable interest to the readership of a number of technical and commercial publications. Your decision to market a particular technology, development of a new technology product, improvements to that technology, the granting of a license—all these and similar events should be considered as candidates for a press release. Second, the price is right! Although no substitute for paid advertising, a series of well-prepared press releases in the technology transfer area can establish an interest in and willingness to work with the various news media in communicating with the outside world.

It is not the purpose of this book to explore the complex world of communications and public relations. Nevertheless, some observations on the development of press releases in the area of technology transfer may be helpful. First, remember that publicity can be a two-edged sword. It may seem to be belaboring the obvious, but check the release for areas of sensitivity *before* it hits the streets. These areas of sensitivity are not limited to your own organization; indeed, the most critical areas may involve a licensee or a partner in a joint venture. Hell hath no fury like a partner in a joint technology transfer effort who feels he has been bypassed on a critical press release. An area requiring even more discretion is that of announcements that may involve the business plans or other proprietary information of a potential client. These may appear rather obvious, but most experienced technology transfer practitioners can cite instances of valid information released prematurely, or from the wrong source, with dire results to the commercial relationship. Second, if the press release involves characteristics of patentable technology, consult your patent attorney. As we noted in Chapter 2, public disclosure can impact patentability. Finally, make it as easy as possible for the recipients of your press release. Prepare it in a style that tells the whole story as concisely as possible. Give a contact for further information. Although it may be revised in the published version, try for a release that can be used as is.

Brochures and Direct Mail

Brochures and direct mail usually bring forth connotations of "junk mail" and various unwanted circulars. Properly prepared and directed, however, technical brochures can be a useful promotional tool in technology transfer. An effective and creative preparation effort is essential, though, to the use of this form of print media. All too often there is an attitude that anything more than the "bare facts" is gilding the lily and obscuring the desired message. The result is often a block flow diagram, some unadorned numerical data, and a strong implied "take-it-or-leave-it" message for the recipient.

The design of a brochure to deliver the desired marketing message is, of course, a function of the kind and number of technology products and the target market. A format that has been found to be quite useful in a wide variety of applications involves the use of a well-designed portfolio folder, into which can be inserted any number of product-specific messages bearing more specific information. The portfolio cover itself usually presents the more generalized institutional messages regarding the organization's capability in technology development and commercialization. In addition, it is in the portfolio cover that a judicial use of color, professional graphics, and institutional logos are most effective in projecting an image of professionalism and competence.

Such a folder can then be used to contain one-page descriptions of various technological products and services. For example, printed and well-designed versions of the black box disclosures discussed in Chapter 5 can be inserted as needed to tailor the mailing to the specific interests of the recipient.

As a general rule, such brochures or other forms of direct mailing efforts are not, in themselves, stand-alone promotional efforts. For maximum effectiveness, they should be used in conjunction with other promotional activities, such as personal contacts or trade shows, which we shall discuss later in this chapter.

Technical Journal Articles

Articles in technical journals are, of course, a form of promotional activity. The primary motivation for the appearance of the article may be debated. Whether it is for publicizing the technology or simply disseminating technical and scientific information, the result is nevertheless a clear message as to the existence of this particular technology product and its potential availability.

Some caveats, however, are in order. First, as discussed in the preceding paragraph, know what your primary motivation for publication is. The choice of preferred journal and readership will most assuredly be different for a commercially motivated promotional effort than for pure technology dissemination. Second, the pen truly is mightier than the sword, or nearly anything else, for that matter. Think carefully about the impact of the publication, particularly as regards premature disclosure. It has been mentioned before, but bears repeating, that prior publication can affect patentability. Moreover, premature disclosure of new technology can impact internal business plans and strategy. Although most organizations have in place a prepublication review procedure, a total reliance on procedures of various sorts may be insufficient to avoid a serious commercial error.

TRADE SHOWS AND EXPOSITIONS

Trade shows, in general, offer a promotional intermediate between broad-based print media approaches and more highly focused personal contact methods. There are indeed an almost infinite variety of trade show promotional opportunities, varying from the enormous West German ACHEMA exposition to far more modest local trade fairs sponsored by city or state chambers of commerce or trade development organizations. In most cases, the sponsoring organization(s) rent booth space in an exhibition hall to the various exhibitors. In return for exhibitors' fees, the sponsors provide space, utilities, some nominal services, and, perhaps most importantly, the publicity and promotion necessary to attract potential buyers and licensees to the event.

Trade shows can be very useful in publicizing and providing initial contacts among a somewhat preselected audience. In most cases, the attendees will be those participating, or at least interested in, a particular industry or segment thereof. However, the percentage of attendees who are decision-makers and/or interested in your particular technological product may be quite small. Usually, the primary objective in a trade-show-based promotional activity in the area of technology transfer is to secure initial contacts for promotional efforts and follow-up.

To this end, then, the planning for participation as an exhibitor in a trade show or exposition will include tactics for "drawing a crowd"—in this case, to your particular booth. It has been clearly demonstrated that even well-designed signs and slogans, pictures of your corporate headquarters, and most institutional advertising posters do not produce overflow crowds. You can use three-dimensional objects that move, that is, working models of the product, if applicable, animated displays, and yes, even the proverbial pretty girl, but remember, however, that your product must compete with the displays. Some have been successful with booths that feature games of skill or chance, counting on these as ice-breakers or conversation starters. There is also the highly subjective, but very significant issue of the balance between ballyhoo and organizational image. That which is natural and effective for one organization may be quite inappropriate for another.

Most exhibitors of technology products have found that proper staffing in the booth is crucial to success. Understaffing in the booth, perhaps as part of an economy measure, is a common problem. Unless your product is very attractive, visitors will not stand in line long to talk to a single attendant who is already tied up with two or three other visitors. Just as important, or perhaps more so, is product knowledgeability on the part of those manning the booth. A plea of ignorance, with an offer to "ask our technical types" and reply later, sends a clear message regarding capability as a supplier of technology.

Two other considerations may be obvious, but are sometimes overlooked or minimized. First, provide a tangible "take-away" message. The brochures discussed earlier in this chapter can be very useful in this regard. These should contain the wherewithal, such as a preaddressed card, for the recipient to request additional information. Second, provide space for registration at the booth to request additional information. In either case, for mailed requests or booth registration, attempt to obtain sufficient information about the visitor's function, job title, and likely role in technology selection to assist in determining appropriate follow-up.

Finally, define a mechanism and response-selection process for follow-up on the contacts thus acquired. Know who is responsible for responding and for determining the type of response, which may vary from minimal non-confidential material to a full-dress state visit. Manning a booth at a trade show can be a grueling experience, and there is often a tendency to slacken one's effort after the conclusion of the show itself. A significant portion of the benefits of participation can be wasted in the absence of prompt, professional follow-up.

PERSONAL CONTACTS

Personal contacts go by a lot of different names—sales calls, customer contacts, client presentations, and many others. Here we are talking about the more highly structured contacts, usually in the potential customer's offices, as opposed to the informal one-on-one encounters over dinner or in some other more relaxed social setting. This is not to downgrade the importance or effectiveness of such informal contacts. However, these are usually tailored to fit the needs of the individuals and issues involved and, as such, are less likely to fit generalized guidelines. Accordingly, we shall confine our discussion to the more organized group contacts.

Of all the forms of technology marketing and sales promotional activity, personal contacts with potential customers are obviously the most focused and specific. The target of the message has been preselected and, it is hoped, the individual recipients of the message have more than a passing interest in the subject being presented. By the same token, however, personal contacts also represent the most time-consuming and resource-intensive form of promotional activity. Because of this, it is critical that there be considerable effort devoted to planning and preparation for personal contact approaches to technology marketing.

Probably the most crucial part of the effort is the selection of the contacts to be made. As we mentioned, the primary virtue of personal contacts is the selectivity thus achieved; poor or careless selection of the contacts to be made can negate any selectivity benefits. Carefully review the material we covered in Chapter 4—Market Research, for approaches and techniques that can serve to improve the selectivity of personal contact promotional activities.

It has been said, and it is undoubtedly true, that common sense is the primary requirement for personal contact sales calls. Nevertheless, how many times have we seen a call fall well short of expectations because of a lack of planning or proper preparation? Let us take a look at some guidelines that can serve to enhance the effectiveness of these contacts.

Do Prepare Professional-Quality Presentations. Value your prospect's time and, more importantly, your own "image" sufficiently to organize and deliver a concise and effective promotional message. Whenever possible and appropriate, prepare and use professional-quality visual aids to get your message across. The use of 35-millimeter slides is usually effective in these presentations, and the necessary equipment is sufficiently portable to be practical. Do not forget the minor details that can make or break a presentation, such as the proper electrical gear to hook up a projector. Remember that current characteristics and plugs will differ as you cross international borders. If your local representative or the potential customer is not supplying the projector, arm yourself with the necessary adapters and extensions. Major surgery on electrical appliances before a room full of potential customers probably does very little for your reputation as a supplier of technology products.

Do Submit An Agenda In Advance. Even though the potential customer may have his own agenda, your input will serve to define the areas of interest and assist in selecting the "guest list." In addition, setting the agenda can be a useful device in maintaining control of the meeting.

Do Research The Customer Beforehand. It may be

obvious, but try to gain an understanding of the customer's major areas of interest. Presumably you have done some of this in the market research phase in identifying potential buyers or licensees for your technology. Nevertheless, the more information you have developed in advance, the more effectively you will be able to direct and focus your presentation. Concentrate your preliminary research efforts on such areas as:

o   Past contacts and transactions with your organization

o   Recent new business, facilities, or organizational announcements

o   Common areas of commercial or technological interests

o   The potential customer's organization and major business areas

o   Who evaluates and selects technology

o   The attendees at the meeting, and their roles in the organization

Obviously, not all this information will be readily available, or even obtainable. The best source may be your own commercial representatives who have had dealings with the potential customer in the past. Trade journals and financial news services can also be useful sources. Above all, do not overlook the straightforward approach of simply asking the prospect. Information gathering does not have to be clandestine to be effective.

To summarize, technology is like most other more tangible products in that it will not sell itself. It may be new, it may be elegant, it may be extraordinarily profitable, but it is unlikely that the world will beat a path to your door, at least until you have delivered some promotional messages regarding the product. We have described a few methods of promoting and advertising your technology product, varying from broad-brush print media approaches to highly selective personal contacts. In the final analysis, it is highly likely that the full range of promotional approaches will be applicable in various stages of the overall marketing plan.

7

Product Pricing

It may be reassuring to students of the "dismal science" of economics to know that many conventional pricing theories are alive and well in the field of technology transfer. There are predictable responses to pricing actions, a fair price is one that offers benefits to both buyer and seller, and negotiation is still required to establish prices. Certainly, there is no reason to believe that pricing of technology products should depart radically from established economic theory.

Nevertheless, there are some unique characteristics to product pricing in the area of technology transfer. First, we are dealing here with a very small number of "sales" of a particular product, even with a very successful licensing program. For major products, such as process licenses, we often see sales of less than 100 units. In addition, technology products are characterized by very high prices, paid in many instances over a long period of time. There are exceptions, of course, but few instances where sale or license of a technology is frequent enough to be considered a candidate for commodity pricing.

Pricing of technology products, such as process technology, also exhibit other unique characteristics. Because the demand for such a product is a function of many factors other than price, there is little, if any, price elasticity. Try as one may, even a significant reduction in price (i.e., license fee) is unlikely to produce an increase in the number of licenses granted, beyond a certain point. Reducing the license fees for a process to produce ethyl alcohol, for example, will not increase the demand for the technology unless there is, indeed, a need for new plants. There is, of course, the possibility of licensing the technology to replace existing manufacturing facilities, but even this is a very limited market.

Another consideration is the fact that major technology products are, in the strictest sense, a performance, rather than a commodity, product. If all process technologies for a given product were functionally the same, price would be a far greater determinant of sales volume. However, when price for a given technology may be only a small percentage of the total revenues at stake for the buyer, other factors, such as licensor's track record, process efficiencies, and technical support capability, can overshadow the impact of pricing actions.

When considering pricing effects, one must also understand some of the legal and political constraints on what one may do in the pricing arena. For example, as we shall discuss later in the chapter on agreements, there are often provisions in license agreements that prohibit lowering the price for one licensee without corresponding reductions for all previous buyers of this particular right to practice a specific technology. These so-called "most favored nation" clauses tend to discourage what we might consider to be market-driven pricing of technology products. There are also, particularly in Third World countries, governmental regulations that place a very restrictive ceiling on the price or royalties one may charge, particularly if the technology is imported. The theory, of course, is that exploitation and profiteering by outside interests is thus avoided, as well as the impact on critical foreign exchange. Without too much effort, one can visualize a situation where legal constraints provide very little latitude for price adjustments, with the floor being set by "most favored nation" clauses in previous agreements, and the ceiling defined by government regulation.

Finally, in most major technology transfer transactions, price is only one of a number of valuable considerations that can be traded off in the course of negotiation toward a commercial agreement. For this reason, it is often very difficult to isolate and compare "pure" pricing effects. As we shall see, price concessions can take many forms, and neither the objective nor the result is to increase the "sales volume."

Having said all this, however, regarding the constraints and complexities of product pricing in technology transfer, it should be emphasized that pricing is an extremely critical element in technology transfer. First, the absolute monetary value of the transactions is usually quite high, amounting to literally millions of dollars or equivalent currency over a period of years. Second, the price of the technology, expressed as a running royalty, can be a significant element of cost or margin for the user's venture employing the technology. Certainly, it can have an impact on the competitiveness and prospects for commercial success. Finally, price can assume an emotional importance all out of proportion to its true economic impact. As one of the most quantifiable elements in the whole transacting, price becomes fair game for much of the negotiating effort and the principal target for those who will be judged by the outcome of those negotiations.

Because of the high profile position of price in technology transfer, it is essential that one have a clear understanding of pricing concepts and guidelines in this area. Although expressed in very quantitative terms, technology product pricing is not a rigorous exercise, but rather a subjective combination of commercial considerations, technical characteristics, political constraints, and human emotion.

PRICING CONCEPTS

There are those whose pricing concept for technology products consists of "... about a penny-a-pound," "...one percent of sales for several years," or "...just a little off the top!". To be fair, there have probably been some technologies licensed with this approach; more often than not, however, with increasing sophistication of buyers worldwide, pricing concepts and bases are required to be thought out, understandable, and defensible. Although a rigorous derivation is seldom required, one must be prepared to develop a rationale, at least, for the proposed fees. While it is not unheard of to work backwards from the desired income to the rationale, one can confidently assume that both will be significant topics in any contract discussions. In addition, in many countries, pricing and other financial terms related to technology transfer will come under close scrutiny by governmental authorities. Let us, therefore, take a closer look at various pricing concepts.

Cost Plus

"Here's how we figure the price to ask for this new process. I've had accounting check the accounting scrolls, and it cost us a cool half-a-million dollars to develop the technology. Add a fifty percent mark-up, and this says we ought to shoot for about $750,000." Perhaps a bit overstated, but fairly representative of the cost-plus approach to technology pricing.

On the face of it, the logic is not bad. You have developed an asset, albeit intangible, and you are trying to realize some financial yield on it. The approach is certainly not unheard of, particularly for some fixed assets, and the price thus developed can without doubt be neatly laid out and derived from an accounting standpoint.

The other side of the coin, however, is that such an approach is obviously cost-, rather than market-, driven. As such, it will seriously underprice the product in most cases. Most assets continue to yield returns, year after year, and such an approach fails to reflect adequately this characteristic. For the most part, this pricing approach has been applied to small technology products which are not a "fit" with the owner's business strategy and where the objective is quick divestiture with minimum follow-up.

Asset Yield

Considering the drawbacks outlined above in the so-called *cost-plus* approach, perhaps we can improve the concept by continuing to recognize the cost of development, but pricing the technology so as to provide the equivalent of a continuing annual yield on the original asset cost. Using the previous example (and an admittedly oversimplified financial approach), let us assume we wish to realize a before-tax yield of thirty percent on the $500,000 it cost us to develop the technology. This would amount to $150,000 per year, or over a ten-year period with the time value of money at ten percent, a present value of approximately $950,000. Certainly, this is a step in the right direction, with a more attractive yield and a recognition of the technology's potential for a continuing stream of earnings over the years. However, there are still some problems. First, it is based on the accounted cost for a unique R&D effort. Even our most enthusiastic R&D friends will usually agree that the correlation between R&D cost and technology product value is, at best, obscure and in many cases non-existent. We have, on the one hand, the programs where years of effort produce little or nothing of commercial value, in spite of very capable management and highly skilled scientists. On the other hand, there are the all-too-infrequent instances where a fortuitous breakthrough will yield spectacular results, completely out of proportion to the resources expended. The point, of course, is that technology product value and development cost correlate poorly and, in some cases, not at all.

The second drawback to the *asset yield* approach to pricing is the same as that for the *cost plus* method. It is cost-derived, and does not reflect the forces at work in the marketplace that determine true product value. In view of this, let us take a look at yet a third pricing approach.

Meet The Competition

On the face of it, this approach sounds good. How can one better recognize the "real world," and the market forces at work therein, than to price one's technology to match the prices currently offered by the competition? Obviously, this would apply only when there are truly competitive products available, so that the *meet the competition* concept will not provide much assistance if you are fortunate enough to be marketing a truly unique product. Aside from that type of happy circumstance, however, *meet the competition* is a pricing concept you will probably encounter during licensing negotiations as a suggestion (or demand!) from your potential licensee. You may have heard the refrain: "You have a wonderful product, we feel very comfortable with you as a licensor. Left to our own devices, we would execute a licensing agreement immediately, but your competition has offered a much lower running royalty which will be very difficult to explain to our management."

Let us take a closer look at the *meet the competition* concept and some of its shortcomings. First, one must indeed meet the competition on straight commodity products, which are functionally identical and offer no product differentiation. Nothing could be farther from a commodity product, however, than a technology product, such as a metallurgical process. Any pricing comparison of technology products must take into account such significant features as ease of operation, raw material and energy usage, and a myriad of other product characteristics that can be quantified and affect product value. Second, even if the inherent product characteristics were identical, other factors, such as experience with the technology, engineering and research support, and training services, can serve to differentiate the product to such an extent that numerical price comparisons are not meaningful. Third, in the absence of the negotiating ploy from a potential licensee cited above, information regarding competitive commercial terms for other transactions involving similar technologies is usually difficult to come by. Trade intelligence can usually be obtained from other agreements involving similar technology, but, as we have noted, comparisons are not meaningful unless all the terms and provisions are available. Such is not usually the case. In view of these considerations, a pricing approach oriented toward meeting the competition is, except in rare instances, not a workable approach.

In view of the above discussion, one may well wonder if there is any workable pricing approach for technology products. Although it is applied and practiced in many different forms, there is a pricing concept that is practical, equitable, and economically sound. It recognizes the realities of the marketplace, it reflects an understanding of the potential licensee's motivation, and it provides a fair return for the owner of the technology. This is known as the *product worth* concept.

Product Worth

Earlier in this discussion, we noted the tendency of some to simplify technology product pricing to some arbitrary percentage of sales over an extended period of years.

Although such an approach is an oversimplification, it does have roots in the *product worth* approach we are now discussing. Basing a product price on product worth to the customer is hardly a novel approach. As a matter of fact, it is in reality merely a restatement of free-market pricing at work in its purest form. As with any concept or guideline, there are exceptions, but these are rather apparent and can usually be accommodated by appropriate modifications in the approach.

Let us now take a closer look at this approach to technology product pricing. First of all, we need to consider the concept of worth to the customer. What is the true worth of a technology product, such as a new process, to a potential licensee? There can be many answers to such a question, depending on how philosophical one wants to be, but a very real and quantifiable response is that worth is closely related to the profit margin a licensee can realize from the right to practice the technology. Although there are some exceptions, such a definition of value covers the majority of technology transfer transactions. Even in non-capitalistic economies, there is a strong awareness of the virtue of competitive economics.

Given this concept of technology product worth, it is not unreasonable to suggest that the price for a technological product should be some portion of that profit margin that represents the worth of the technology to the licensee. What portion? Here we move rapidly from the quantitative realm to a more judgmental area. This issue has been explored in excruciating detail through countless negotiations and in scholarly discussions, usually with results based on compromise and commercial reality. In general, the pattern has been general agreement that the owner or licensor of the technology should receive less than half the net profit margin from the venture, expressed as a percent of sales. This reflects the higher level of financial risk undertaken by the licensee, as well as the responsibility for other costs associated with the venture. At the other end of the scale, the licensor's share is seldom less than ten percent of the net profit margin, again expressed as a percent of sales from the venture. Where in this range of ten to fifty percent of net profit should the licensor's share be fixed? Well, as the saying goes, it depends. For unique one-of-a-kind technologies, a share of twenty-five to fifty percent is probably appropriate. For those technologies falling into the "me too" category, a ten to twenty percent share is as much as can be expected in view of competitive pressures from other licensors.

Although derived as a share of net profits, the licensor's share, or running royalty, is usually expressed as a percent of net sales from the licensee's facility. If, for example, the technology product is a process that produces a commodity petrochemical, the net profit for the operator or licensee may amount to approximately ten percent of sales. If the technology is a reasonably competitive one, a licensor's share of perhaps twenty-five percent of the profit is not unrealistic. The running royalty would then be expressed as twenty-five percent of the ten percent margin, or two-and-a-half percent of net sales.

Although the approach in this pricing method is numerical and quantitative, the application still involves subjective judgment, particularly as regards the split of profit between licensor and licensee. As a matter of fact, it is this combination of quantitative and qualitative factors that makes it a preferred method of pricing technology products.

As a matter of interest, a review of the application of this concept to past technology licensing transactions indicates that the method results in license fees or running royalties that vary from one to four percent of net sales for commodity-type petrochemicals. For specialties and pharmaceuticals, for example, the running royalties as a percent of sales are higher, reflecting of course, the correspondingly higher profit margins usually realized in these product areas.

There is one message that comes through very clearly in the application of the *product worth* method in particular, and in other areas of technology transfer in general. That is the desirability (necessity may be more accurate) of understanding as much about the potential customer's economic situation as possible. Granted that the prospective licensee is not likely to disclose the financial analyses of his venture, it is nevertheless possible to scope out a reasonable approximation of the profit margin and critical cost elements of the proposed facility. Believe it or not, knowledge is power, at least at the negotiating table.

Although it is the preferred pricing method, an understanding of the limitations of the *product worth* concept is essential. First, it is intended as a means of establishing a reasonable price negotiating range, not as a rigorous derivation of "the price." Second, some economic systems do not explicitly recognize profit in the capitalistic sense, and ideological successes in the implementation of a five-year plan are not easily translatable to the calculation of license fees. Nevertheless, the concept can still be applied internally using world prices, and the result then converted to a different form, such as a lump-sum payment, by discounting to a present value at an appropriate interest rate. Third, it may seem at first glance to be inapplicable to the situation wherein the new technology is being used to replace a less economical facility, and there are no net additional sales. Here again, however, we can establish *worth* to the licensee which, in this case, is the reduction in manufacturing cost he can realize by having the right to practice the new technology. A reasonable license fee can still be developed using the concept.

Nearly as important as pricing concepts, albeit in a negative sense, are some of the myths that have grown up around the area of technology product pricing. Usually quoted without attribution, they nevertheless appear to persist and survive in an almost clandestine manner. Because these myths do significantly affect the development and negotiation of technology product prices, a recognition and understanding of them is a useful adjunct to the general area of product pricing.

Pricing Myths

"Licensing income is the icing on the cake." Commonly heard in discussions of product pricing, this statement is also known as the "something-for-nothing" syndrome. Usually the statement reflects a feeling that technology transfer is not really a business, that licensing income is really a stroke of good fortune, and that one should be profoundly grateful for any income, regardless of how small it might be. The corollary, of course, is that little or no effort should be expended in producing this insignificant flow of income.

The significant loss of income and asset value is one very real consequence of this particular myth. However, the real danger in this outlook lies in the temptation to carry out the technology transfer in a slipshod, minimum-effort manner. The result can be time-consuming fix-ups, embarrassing and costly litigation, and permanent damage to one's position as a reliable supplier of technology products. Technology transfer income may appear to be a bonus or "icing on the cake," but to avoid serious problems, technology must be handled like any other component of the product line, with tender loving care and full resource support.

"The customer will pay what it would otherwise cost him to develop the technology for himself." This myth is somewhat similar to our *cost recovery* pricing concept discussed previously, except that here we are talking about the customer's hypothetical cost of development, as opposed to our own real costs. Even those R&D managers with extensive experience and expertise in program budgeting can testify as to the difficulty of estimating such development costs. If, in spite of the obvious difficulties, one could estimate the costs of such a program, there still remains the fact that such a cost, accurate or not, has absolutely no relationship to the product worth. It would indeed be a very neat and tidy argument for the negotiating table to equate the worth of the technology to development cost; unfortunately, the crystal ball to quantify the argument does not exist.

"Any product can be priced to sell." Hark back, if you will, to our discussion on price elasticity in the early part of this chapter. Price is not, in most cases, the prime determinant of sales volume. The demand for "big-ticket" technology products, such as process know how, is created not by price, but rather by the need for the tangible product produced by that process. It is probably fair to say that even a royalty-free grant would be a tough sell for a process technology to produce an unwanted or unneeded product. This should be an obvious point, but truly one that is often overlooked in initial pricing discussions on a new technology product. The familiar refrain of "Cut the price, and we'll make it up on the volume" is indeed applicable with certain tangible products. Nevertheless, this particular myth has been responsible for underpricing and failure to realize the full financial potential of a number of new products. In the field of technology transfer, initial pricing and price trend expectations are critical; mythical pricing premises can lock the licensor into very undesirable situation.

"The licensee's future competitive position is his problem." We have discussed, in regard to some of the other technology transfer myths, the generic problem of underpricing a technology. There can be, believe it or not, also problems with overpricing the technology. At first glance, one might anticipate that the problem would be limited to lack of sales or being priced out of the market. Although this is indeed one possible result of overpricing, there is also another drawback. This involves the impact of high running royalty payments on the competitive position of the licensee and the potentially harmful effect of deteriorating economics on the future stream of royalty income to the licensor.

Note that we are not suggesting that licensing be a philanthropic activity. The main message is that a licensing agreement is, in a sense, a de facto partnership. Both parties have an incentive, from a pure self-interest standpoint, if nothing else, to make the arrangement work. The licensee's motivation is apparent, with his whole venture at stake. The licensor, however, is also at risk with regard to his future licensing income from the venture and, more importantly, his reputation as a reliable supplier of technology products. An enlightened self-interest in the licensor's future competitive position is one of the reasons for periodic information-exchange meetings of a licensor with its licensees, such as Bayer on vinyl acetate and Phillips on polyethylene. In general, there is a clear incentive for the licensor to regard the licensee as an important partner in a joint technology venture, and to price accordingly.

FORMS OF PAYMENT

Thus far, we have discussed pricing concepts in terms of one form of payment, that of so-called running royalties which are based on the amount of product produced and are payable over a period of years. Other minor references have been made to various forms of lump-sum payments. Although a determination of the total income to be realized from the technology transfer is quite important, the form and pattern of payments are also critical considerations in technology product pricing.

It is probably apparent that a running royalty, paid on a quarterly or even an annual basis, has the virtue of better matching the cash flow from the enterprise using the technology. The financial obligations are created at the same rate the venture produces income for its owner, the licensee. From the licensor's standpoint, the running royalty creates a continuing stream of income to subsidize future technology development and marketing efforts. In the licensor's own organization, this continuing income can serve to alleviate some of the "What have you done for me lately?" response to proposals for licensing support. All in all, for most technology transfers, a running royalty form of payment, based on a percent of sales, would appear to be a fair and equitable method of pricing.

One drawback to a "pure" running royalty payment form is the shortage of up-front money, first, to cover expenses associated with the technology transfer and second, to ensure the licensee's commitment to the venture in advance of other obligations, such as capital expenditures. It is true, of course, that many of the technology transfer expenses, such as disclosures and other technical support, are covered on a per diem technical support agreement. Nevertheless, there are others, such as business development and administrative activities, that must be absorbed by the licensor. In order to deal with this problem, the pure running royalty form of payment is frequently modified to include a lump sum *plus* a running royalty. The amount of the lump sum payable in advance of actual operations is a judgment call, usually modified and adjusted during negotiations. The maximum up-front payment is sometimes calculated as the equivalent of a year's running royalty at capacity operation, and is often scaled down from there.

It can be seen that all of the above approaches have a common characteristic, namely, the administrative cost associated with collection, accounting, and sometimes auditing of the quarterly or annual remittances of royalties. In addition, there can be significant risk in some areas of the world in reliance on extended payments over a long period of time, e.g., governments and economic systems can change, and attitudes toward the use of foreign exchange and remittances to foreign licensors are subject to drastic revision. These considerations and others can involve, in some cases, unacceptable levels of business risk. For this reason, the running royalty form of payment, even with an advance up-front payment, is sometimes modified to one fixed-price, lump-sum payment, which may be paid in installments over the period from license execution to satisfactory operation at guarantee levels.

The method for calculating the lump-sum payment can vary, but the basis is usually the running royalty developed in the manner described earlier. An assumption is made regarding the operating rate of the facility and the net sales price over the term of the royalty period to calculate the royalty payments each year. Using an assumed factor for the time value of money (say, ten or twelve percent), these payments are discounted and summed to yield a present value, which is then used as a first pass at the lump-sum payment. There are refinements to this procedure that can be applied, but are usually not warranted, considering the number of assumptions and predictions already inherent in the method.

A principal application of the lump-sum approach to pricing occurs when technology fees are combined with payments for the facility itself, as in the case of a joint venture of a licensor with an engineering contractor. In this case, the engineering contractor submits a price which includes both the facility cost and the license fee. The price is then paid in installments over the period of time from contract execution through facility commissioning, start-up, and guarantee runs. Although such a package alleviates the long-term accounting and auditing problems inherent in a running royalty arrangement, care must be taken to ensure appropriate tax treatment by the licensee's government, inasmuch as tax rates for know-how and technology fees sometimes differ from those for equipment and other hard goods.

ALTERNATE FORMS OF PAYMENT

"I don't have any problem with technology licensing, so long as we get our money back to the United States in cold hard U.S. dollars." Undoubtedly paraphrased a bit, but fairly representative of the outlook of some organizational managements regarding payment of license and related technology fees. Here again, there is no absolutely right or wrong answer to the question of flexibility in payment forms. Rather, the need is for a careful understanding of what is right for one's own organization and of the trade-offs inherent in one's position on flexibility and creativity in payment forms. If the technology product is being marketed only in the U.S., or if the technology is unique with little or no real competition, the rather dogmatic approach outlined in the quotation above may be a viable one. On the other hand, if there is significant competition, especially by experienced licensors with trading capability, the marketing effort, even for a very attractive technology, may be an uphill battle. In most cases, it is possible to exercise flexibility and creativity in structuring financial terms without violating good business practices and sound toward technology transfer as a business.

Let us now take a brief, but closer, look at a few alternative forms of payment for license and technology fees.

ALTERNATE CURRENCIES

Requests for payment of license fees and related costs in something other than the licensee's own currency are quite common in international technology transfer. There are usually two somewhat related reasons for these requests. First, in many countries, particularly Third World nations, foreign exchange is a very critical issue. Because of trade balances and national policy, it may be very difficult to come up with the needed hard currency, such as yen, marks, or dollars to make the necessary payments of technology and service fees. A processing facility serving domestic needs will not alleviate the foreign exchange problem, even when operating successfully at better than nameplate capacity. A would-be licensor who is able to address this problem in a creative fashion has an obvious competitive advantage.

The second currency-related problem is not limited to developing nations with foreign exchange difficulties. Rather, it involves the understandable reluctance of any licensee organization to incur future financial obligations in something other than its own currency. In the early 1970's, for example, a U.S. firm executed a license agreement calling for payment to a German licensor of a license fee over a period of approximately three years. At the time of the agreement, the exchange rate was about four marks to the dollar. Over the payment period, however, the mark increased in value against the dollar to an exchange rate of about two-to-one, resulting in an increase of three to four million dollars over the intended payments by the licensee under the agreement.

A comprehensive discussion of international currency management and strategy is beyond the scope of this book. However, a brief look at several approaches that have been used in dealing with this alternate currency problem may be useful in generating some conceptual approaches by the reader for specific situations.

First, consider the soft currency/foreign exchange problem. In some instances, the licensor has accepted the licensee's currency, at least in part, and used it to purchase needed raw materials or natural resources, sometimes at a very favorable price. Obviously, this approach is not risk-free and can sometimes entail tedious bureaucratic procedures to implement. Nevertheless, even if applied only to a portion of the total fees, such an approach can yield a competitive edge and some profitable trading opportunities.

Second, with regard to both the soft currency and the more general exchange rate problem, another approach has been to design a composite currency unit, comprising varying portions of different convertible currencies, such as yen, marks, francs, dollars, etc. Such a payment form can provide needed latitude to the licensee, who may then be able to make use of more favorable trace balances with other countries in order to meet the financial obligations of the license agreement. In addition, aside from the soft currency problem, a mixed or composite currency arrangement can alleviate some of the risk associated with future exchange rate fluctuations found in most international technology transfer transactions.

Finally, and perhaps obviously, there is the use of currency futures to hedge the risk inherent in future financial obligations denominated in currency other than one's own. Such an approach is most applicable to those transactions that involve relatively large sums over a short time period. Many organizations involved in technology transfer have international currency specialists or departments; others make use of outside resources, such as banks and other financial institutions.

PRODUCT TAKE

Any licensing manager who has dealt with technology transfer activities in the so-called Third World countries is familiar with the proposal to pay license fees by delivering product produced with the licensed technology in lieu of monetary payments. Such an arrangement is obviously attractive to the licensee, providing a more favorable operating rate for the new facility, as well as conserving scarce foreign exchange. Some licensors will reject product take arrangements out-of-hand, contending that the marketing and logistical problems associated with the disposal of product accepted in lieu of royalty payments outweigh any benefits. Other licensors, however, and particularly those associated with trading companies, are prepared to deal with product take proposals, and to implement such arrangements when conditions are favorable.

In most cases, product take will only be used to offset a portion of the monetary payments required under the license agreement. If the product can be obtained at a value equivalent to its *incremental* cost of production, as opposed to a value reflecting either full-absorbed cost or sales value, there may be an opportunity for a mutually beneficial arrangement. In most cases, the licensor of the technology will be a participant in the product market and may have good access to geographically feasible customer outlets. Particular care must be taken to address the questions relating to distribution cost and duties. Although product take arrangements are not by any means always the right approach, a willingness to consider and explore such approaches can enhance, under the right circumstances, the profit potential of the technology transfer agreement.

Product take is actually one very specific form of a general international commerce activity known as *countertrade*. Countertrade, in its most general sense, involves the exchange of goods and services, including technology, instead of straightforward monetary transactions. Suppliers of commercial aircraft, for example, have become proficient in structuring large sales agreements to include not only cash, but also hams, packaged tours, and other indigenous products and services. In certain instances, some form of countertrade can be a valuable adjunct to technology transfer transactions.

EQUITY OR OWNERSHIP POSITION

On occasion, a potential licensee will propose that all or a portion of the technology fees be paid, in effect, with a share in the ownership of the facility using the technology. The rationale for such a proposal, in addition to foreign exchange problems, is that such an arrangement ensures a high level of commitment on the part of the licensor to the success of the venture. This approach, in reality, converts a technology transfer transaction into a joint venture.

Most licensors will avoid such an arrangement, unless there are some very special advantageous features. Seldom does the equity position offer any control over the decision-making in the venture. Repatriation of any profits from the venture is often difficult, if not impossible. Future liquidation of the position usually entails a significant financial penalty. An equity position as an alternate form of payment may be attractive under some circumstances, but is, in general, one of the least attractive alternate forms of payment.

To review briefly, a key element in any form of technology transfer transaction is product pricing. Although expressed in a very quantitative form, there are a number of subjective judgments involved in technology price development. The recommended pricing concept involves a share of the estimated worth of the technology to a potential licensee and a running royalty over a period of years. There are, however, circumstances where other forms of payment such as a lump sum or a variation thereof, are preferable. Finally, a willingness to exhibit some creativity in considering different compensation arrangements, such as alternate currencies, product take, and other forms of countertrade, can considerably enhance the marketing effort and possibly increase the total profitability of the technology transfer effort.

8

Sales and Negotiations

In previous chapters, we have explored various marketing activities. These include the decision to market technological products, the market research required to establish potential customers or licensees, the packaging of the product for marketing, the promotional effort required, and product pricing approaches. Although all licensing efforts will not include all steps in this sequence, such steps nevertheless are representative of the process required to reach the target of a technology sale or license. In a generic sense, all these activities could be considered part of the "selling effort." However, in this chapter, selling or sales effort is used in a somewhat more restrictive sense to define the activity required to progress from an identified, interested prospect to an executed license agreement.

There is, however, one other very vital selling effort that may have preceded those described above. That is the internal selling effort that may be required to convince one's own management that the marketing of technology products can be an effective element of the overall business strategy. Reference is made to Chapters 1 and 2, wherein we presented an overview of some of the selling effort required to overcome internal resistance to the concept of parting with the "crown jewels" of technology via an aggressive licensing program. For the purpose of this chapter, it is assumed that an effective internal selling job has been implemented, and that the focus is on external customer sales.

Let us, therefore, assume that we have done a thorough job of market research, marketing package preparation, and product promotion. As a result of our efforts, we receive an inquiry from a prospective licensee. In the inquiry, our prospect requests a detailed technical disclosure, commercial terms and a plant visit to our manufacturing facility using the technology. How should we respond, what actions should be undertaken to follow up on the inquiry, and what questions need to be addressed? In addition, if we follow through properly, the preliminary response to the inquiry can lead ultimately to serious negotiations on a license agreement, and we need to consider how we best prepare for and implement this critical activity.

Although there is no such thing as a "standard" sales effort in technology transfer, most sales activities entail addressing some version of the following issues:

o   Initial response

o   Preparation and research

o   First meeting

o   Secrecy issues

o   Negotiating approach

Although this list is by no means a comprehensive one, most of the significant activities in sales or technological products fit into one or more of these five categories. Let us now take a closer look at these issues and their impact on the success or failure of the selling effort.

INITIAL RESPONSE

As mentioned previously, let us assume that we have received an inquiry from a reputable organization requesting a complete technical disclosure, commercial terms, and a visit to our manufacturing facility that uses the technology. Inasmuch as "the customer is always right," do we therefore gear up a massive effort to produce full disclosures, a complete licensing agreement, and a "no holds barred" inspection tour of our manufacturing facilities? The answer, as you might imagine, is absolutely not. A full technical disclosure without the benefit of a clear understanding of the potential client's capacity requirements, energy constraints, staffing plans, and technological capability, to mention but a few of the critical factors, would be an exercise in futility. A complete license agreement without a good picture of the client's financial situation, governmental regulations, and foreign exchange outlook would serve only to waste the client's time, as well as the efforts of our attorney. (One exception to this observation could be the frequently licensed process, such as drying systems or acid-gas removal systems, for which terms have been well established within a particular market.) Finally, a visit to our manufacturing facility at this preliminary stage could result in a premature disclosure of vital aspects of the technology with no compensation for the inadvertent technology transfer. It should be mentioned, moreover, that such overkill in the initial inquiry usually comes from less-sophisticated organizations, who may be oriented toward requests for proposals on tangible, commodity-type products. In most cases, initial inquiries from organizations knowledgeable in licensing and technology will follow a more traditional pattern.

Be that as it may, however, we have this all-encompassing inquiry in hand, and we need to respond in a timely, professional manner. Let us consider some approaches.

With regard to the request for a full technical disclosure, respond with the most comprehensive non-confidential disclosure available. The disclosure should be personalized to the greatest extent possible, with references to the inquirer in the disclosure itself and, where practical, modifications to fit the specific inquiry. Use the response as a vehicle to highlight the competitive advantages of the technology, such as productivity, efficiency, or energy utilization. A non-confidential technology disclosure can be useful and informative without jeopardizing the confidential or proprietary aspects of the technology. It can also serve as an effective selling device, so long as it does not appear to have originated from stacks of identical material in some impersonal mailing room.

So far as the request for comprehensive commercial terms is concerned, the response will depend on the type of technology in question. With less complex, standardized, repetitive technologies, there may indeed be a standardized "boiler-plate" version of a licensing agreement available, which can be forwarded with appropriate blanks for information unique and specific to a particular licensee, such as desired capacity, site conditions, and so forth. In many cases, however, the technology will be a rarely licensed product, with no feasible standard contract or terms available. In this situation, avoid trying to outguess the inquirer's needs and requirements by formulating financial terms at this preliminary stage. Use the reply instead to elicit additional information from the inquirer that can serve to define better his specific needs.

In some instances, the rationale by the inquirer for requesting such comprehensive technical and commercial information is that a decision has been made to proceed with the project (finally!) and that all information must be in hand within, say, 60 days so that a final, irrevocable decision can be made at that time. Obviously, one must make a judgment regarding the credibility of the source, as well as the trade-offs involved in responding on a crash basis. If the decision is to reply in accordance with the inquirer's schedule, significant contingency factors should be built in to both technical and financial responses. Use conservative productivity and raw material efficiency factors, and structure license fees and running royalties with a clear understanding that licensing income can only decrease, never improve, from that point on. Except in very unusual circumstances, however, one should resist the temptation of being stampeded into imprudent disclosures because of unreasonable time constraints. The missed commercial opportunity is usually more than offset by avoidance of an ill-conceived, embarrassing, and costly commercial venture.

In addition to these guidelines on requests for technology disclosures and commercial terms, we still must deal with the inquirer's interest in visiting the facilities that employ our technology. First of all, include as much non-confidential information as possible in your reply to the inquiry. Consider the reply as a special opportunity to highlight the track record of the facility, its capacity and on-stream time, and its safety record. Point out its advantages over competing technology products, as well as improvements that have been made in the facilities.

Such a reply will not substitute for an actual visit to the facilities by the potential licensee, so that plans should be made for the visit itself as an integral part of the selling effort. Include in the reply an execution copy of a secrecy agreement to cover a future plant visit. Schedule the visit as a part of the first meeting with the potential customer.

For the visit itself, careful planning is a must. The execution of a secrecy agreement by the visitor does not entitle him to a full disclosure of all the details of the technology. For this first visit, the principal objective is usually to send a message regarding commercial reality, ease of operation, and the capability of the licensor as a resource in technology transfer. It is not intended, at this stage of the sales effort, as a means of transferring the technology and know-how. Because of this, take some time and effort to brief very carefully the plant personnel who may come in contact with the visitors as to the objectives of the visit. Be sure all involved have a clear understanding of what the visitors can hear and see during their visits, as well as any commercially or technically sensitive areas. All too often, the assumption that all communications can be "filtered" through the licensing manager or other commercial escort becomes very difficult to implement, especially with large groups of visitors.

From the foregoing discussion, it can be seen that the initial response to a potential customer leads quickly to a consideration of needed preparation for the first meeting with our prospect. It is these preparations that can play a vital role in setting the tone for the whole sales and technology transfer effort.

PREPARATION AND RESEARCH

The homework associated with the preparation for the first-face to-face meeting with a potential customer usually falls into two major categories: that needed to define the product status and position within our own organization, and that required to understand better the situation of the potential licensor. Both need to be considered as part of an effective sales effort.

Internal Preparation

Thus far, our emphasis has been on marketing of technology products for which there has been an overt market development effort. The pros and cons of licensing have been debated and analyzed (sometimes ad infinitum), market research and promotional activity have been carried out, and the organization is generally identified with the effort. However, it is not uncommon to receive an inquiry for a technology product for which there has been no preparation for licensing. In this case, internal preparation for the first meeting becomes quite important.

Some of the questions that should be addressed in advance are as follows:

o Do we have an organizational commitment to license this technology?

o Do we have the resources required to transfer the technology, such as

(1) experienced operating personnel?
(2) capable engineering and R&D staff?
(3) adequate documentation of the technology?

o Do we know what financial incentive we need to provide an adequate return on the resources?

o If so, is that financial expectation reasonable, considering the competitive situation and the economics of the product made by using the technology?

Although it is highly unlikely that any sort of commercial commitment will be made or expected at the first meeting, a lack of clear-cut direction and understanding on the above points can raise some serious concerns in the mind of the potential customer as to the wisdom of proceeding further with the effort.

External Preparation

Irrespective of the level of and need for internal preparation, external preparation that will yield a better understanding of the potential customer's plans, organizational background, and general business situation will serve to enhance considerably the chances for commercial success.

Check your own organization for other commercial interfaces. Is there a customer or supplier relationship, and in what product areas? Is the relationship currently satisfactory, or are there some sensitive areas where one should tread lightly? Many times these other commercial relationships are used by one of the parties to the transaction as leverage in future negotiations. There may be a strong incentive, for example, to provide certain concessions to a potential licensee who is also a valued customer for other products, either technological or tangible. An advance understanding of this eventuality can be very useful in tailoring the sales effort to fit.

Investigate also the inquirer's current and past activities in this particular technology area. Does he now use competing technology products, whether his own or acquired from others? What is his level of technological sophistication, and what resources does he have to facilitate the technology transfer? The latter can be a key question, inasmuch as the critical issues in technology transfer can vary considerably, depending on the technological expertise of the potential licensee. A less technologically sophisticated prospect presumably will require a greater selling emphasis in the area of engineering assistance and review, commissioning and start-up help, and past performance in the technology transfer area.

Financially related information, while sometimes not easily available, can also be quite useful in preparing for the first meeting. Credit worthiness is an obvious consideration, but more subtle financial aspects, such as economic criteria and access to financing, can be useful input in structuring financial terms.

In general, the better the picture that can be developed of the potential licensee, the more effective will be the selling effort presented in the first meeting. Time does not always permit exhaustive research prior to the meeting, but any information that can be obtained and applied will serve to enhance the selling effort.

FIRST MEETING

To the extent possible, the licensor of the technology should try to ensure management and control of the meeting. This is not to say that the meeting should not be responsive to the needs and objectives of the potential licensee; indeed, the meeting should reflect, within the constraints of proprietary information, the content of the initial inquiry. Nevertheless. control of the meeting location, agenda, timing, and presentation approach by the licensor can serve to convey an impression of capability, as well as ensure satisfactory direction of the marketing effort. Nothing detracts more from an image as a reliable supplier of technology than a meeting agenda that can be paraphrased as "Welcome, and what would you good people like to talk about today?"

With regard to location, the first meeting should usually be held at the licensor's offices. In many instances, this will be the only logical location because of planned visits to the licensor's laboratories or manufacturing facilities. Even without these planned visits, a meeting at the licensor's location offers advantages in terms of access to experienced technical and commercial personnel that may be required as part of the selling effort.

The agenda for the meeting will be a key element in the effectiveness of the sales effort. Not only does it define the material to be covered, but it can also serve as a control device for the selling effort, ensuring emphasis in the appropriate areas. Wherever possible, the agenda should be reviewed in advance by the potential licensee to ensure consistency with his needs. In structuring the agenda, separation of technical and commercial sections can be beneficial in arranging attendance and representation by specialized functions. With regard to attendance, make a valiant effort to hold the attendees down to those actually needed to cover the material, realizing that this may be an unrealistic expectation in some organizations.

For the presentations themselves, it may be obvious, but professionalism is a must. Organized and preplanned remarks, clearly understood objectives, and effective visual aids all contribute to the effectiveness of the selling effort. This may appear at first to be belaboring the obvious, but most of us have seen some fairly amateurish performances under the guise of "an informal, low-key approach."

It is unrealistic to expect that the meeting will rigorously follow some structured script prepared in advance of the meeting. There must be room for "give and take," with allowance for deviations from any prepared program. Nevertheless, one should be alert for impromptu or off-the-cuff remarks or claims that could come back in a misinterpreted version during some future meeting. In some instances, a remark intended in a humorous vein will be misunderstood and serve as a diversion from the main focus of the meeting. Because of this, careful monitoring of the discussion is desirable, with clarifying or explanatory remarks as appropriate. Particular care should be taken with side discussions, where there may be a significant opportunity for isolated misunderstandings.

In many instances, there will be a language barrier associated with the meeting because of the different nationalities involved. In these cases, particular preparation efforts and care in the conduct of the meeting are needed. Where there is no interpreter and the visitors are expressing themselves in your language, do not assume that a head nod or a smile denotes comprehension; it may, in fact, indicate nothing more than courtesy and a friendly attitude. It may be useful to follow a statement with a question as to your listener's feelings on the subject, so as to ensure comprehension on critical issues.

In those cases where an interpreter is being used, check the participants for their experience in making presentations through interpreters. Where practical, give the less experienced participants a chance to "rehearse" their presentations with an interpreter, to get a feel for timing, feedback response, and pacing. Be very certain that everyone understands that a language difference is no guarantee that a confidential statement made in your native language will not be understood and filed away for future reference by your visitors. In the absence of evidence to the contrary, assume that any audible statement will be understood by all.

SECRECY ISSUES

Many an initial contact or first meeting has been significantly complicated by questions of secrecy and related disclosure problems. In many instances, the prospective licensee is very anxious to obtain as complete a disclosure as possible, so as to facilitate a thorough evaluation of the technology. The licensor, on the other hand, is not inclined to completely "bare his soul" without some tangible recognition of the value of the technology and some safeguards against uncontrolled disclosure. With sophisticated potential licensees, the problem is usually not as severe, because of their background and experience in technology transfer. With others, however, the secrecy issue can be somewhat more difficult to resolve.

In many cases, the transmittal of an appropriate secrecy agreement for execution in advance of the meeting can serve to condition the prospective licensee to the problem and the proposed approach. In some cases, for the secrecy agreement to be well received, it should be accompanied by an explanation outlining proprietary concerns and defining the disclosure plans and sequence. As is the case in many areas of commerce, careful efforts at an early stage to clarify reasons for a position taken can serve to avoid or alleviate undesirable consequences.

A common secrecy issue involves a potential licensee who may be carrying out, in his own laboratories, R&D explorations in the same technology area. Understandably, he may be quite reluctant to execute a secrecy agreement, inasmuch as the field of the agreement might cover subsequent developments that he might make on his own. Nevertheless, he may be quite anxious to obtain sufficient information regarding your technology to permit evaluation and comparison with his own version of the technology. The licensor also has a strong incentive to work out an arrangement to permit the evaluation, inasmuch as this is obviously a serious prospect. The challenge is to devise a method that will permit the evaluation without compromising either party's position.

One approach that is often taken to deal with this problem is to restrict an individual or organizational unit from a secrecy standpoint, and then proceed to make the disclosure for evaluation. In some cases, a knowledgeable individual, perhaps close to retirement, in the potential licensee's organization receives the disclosure but is contractually restrained from subsequent activities in the subject field of technology. In addition, he or she is prohibited by personal secrecy provisions from disclosing the technology to others within the organization. This is sometimes referred to as the "Typhoid Mary" approach, because the recipient of the sensitive information is, in effect, quarantined to prevent unauthorized and potentially damaging disclosures of the technology.

When the licensor is concerned about the use of a member of the potential licensee's organization as the "Typhoid Mary," a third party, such as a consultant or contractor, is sometimes used. In this instance, the third party is retained to receive the disclosure from the licensor under strict secrecy terms. Using the disclosure, the third party proceeds to carry out an economic evaluation of the technology. The evaluations are carried out in accordance with guidelines and ground rules provided by the potential licensee, and the final report, comprising commercial rather than technical data, is reviewed by the licensor to ensure protection of proprietary information.

NEGOTIATING APPROACHES

It is not the objective of this book to provide an in-depth analysis of the art of negotiation. There are a considerable number of useful books on the subject, as well as a plentiful supply of seminars and how-to courses in this area. There is also a considerable body of thought that there is, in reality, no substitute for actual experience in the negotiation of commercial agreements. Nevertheless, a closer look at the negotiating phase of technology transfer can be useful in maximizing the effectiveness of the sales effort.

Consider some generalized guidelines. First, rest assured that there is, in reality, no one "right" negotiating style. Aggressive, deferential, formal, casual,--all have been used effectively, depending on the understanding gained of the other party, and the circumstances applying to that particular stage of the effort. The "right" negotiating style is that which works for the particular issues and parties involved, and this can change as the negotiations progress. Keep in mind that among the most useful guidelines to negotiating styles are the reaction and responses to the other party. A conscious effort on your part, or that of a colleague, to "read" the reaction of the other parties and to analyze them after each session can pay dividends in effectiveness.

Second, be aware of your negotiating approach and goals early in the sales effort. It may seem premature to concern yourself, before the first meeting with a potential licensee, with the kind of commercial arrangements you hope to realize ultimately. Nevertheless, a common understanding within your own organization of the commercial goals can prevent inadvertent misstatements and commercial blunders in early discussions. Although a complete draft of a licensing agreement is not usually a practical approach, an outline of commercial terms, or Heads of Agreement, summarizing initial positions on such issues as financial terms, exclusivity, transfer packages, technical support, guarantees, and grantbacks (rights to subsequent technical developments by licensee) can serve to provide common understanding and guidelines for all involved.

Third, once commercial terms (or an outline thereof) have been submitted, there will quite likely be counterproposals as part of the negotiating process. In preparation for this, one should have developed an understanding of acceptable trade-offs that might be used in the negotiations. As has been cited on numerous occasions, a key consideration is to identify elements of commercial terms that may be of more value (real or perceived) to one party than to the other. Depending on circumstance and the parties involved, such considerations as geographical exclusivity, additional capacity at reduced royalty rates, or adjustments in payment format ("up-front" versus running royalty) can serve as effective trade-offs in the negotiations.

Fourth, when concessions or trade-offs are made, take the initiative of redrafting the agreement or outline of terms yourself. Although this may involve some rather primitive or makeshift approaches when the negotiations are carried out at a remote site, the additional control achieved is well worth the extra effort and aggravation. A prompt redrafting effort can also serve to preserve the momentum of the negotiations.

Fifth, try to understand, as early as possible in the discussions, the decision-making process in the potential licensee's organization. Find out, for example, if those present for the other party can make decisions, and, if so, at what level. In general, a big difference in on-site decision-making power between the parties to a negotiation can make for difficult discussions. Although it is obviously impossible to dictate to the other party their representation in the discussion, an awareness of imbalances can be helpful in avoiding hasty responses to concessions that may be repudiated later.

Sixth, gain an understanding of the national laws and policies regarding technology transfer under which the potential licensee must operate. A favorite and usually effective approach of the potential licensee is to contend that his government requires more favorable terms as a matter of national policy. Indeed, in many cases this is true, particularly in Third World countries. Nevertheless, it is critical that these restrictions be understood at an early stage in order to structure a realistic set of commercial terms, and also to understand the extent of the review process. It is also important to explore, through local contacts, the level of enforcement for these restrictive provisions. For technological products that are deemed necessary for trade or policy objectives, there may be latitude and flexibility that are not clearly reflected in the written version of the regulations.

THIRD PARTY SALES EFFORTS

Thus far, we have confined our discussions to in-house sales efforts carried out by commercial and technical personnel of the licensor. Another approach that is sometimes used, particularly when resources are limited, is that of using a third party for the selling effort. The third party can be an engineering contractor, a consultant, or even, in some cases, a previous licensee with rights to sublicense. Properly planned and structured, a third party sales arrangement can indeed extend scarce commercial and technical resources and, in some cases, have a synergistic effect on the total marketing effort. However, as discussed in Chapter 4—Market Research, considerable care must be exercised in the development and execution of such arrangements.

First of all, take sufficient time and effort to ensure proper selection of the sales representative. The contractual relationship probably will last a long time and involve significant financial stakes. Find out as much as possible about his track record, particularly in the area of technology transfer. Be able to identify exactly what the representative brings to the effort, and how this can enhance the chances for success. Are his strengths commercial, technical, or financial? Do these qualities mesh with your own strengths, or do they overlap?

Understand the potential partner's motivation. It can be quite different, as in the case of an engineering contractor's aspirations to sell engineering services, without conflicting with your own objectives. Nevertheless, a clear understanding of the partner's driving force, however different, is essential to a workable partnership. A previous licensee, for example, would be understandably reluctant to support licensing a potential competitor in the same general geographical area.

Take care to spell out the respective accountabilities in the joint sales effort. Who, for example, is authorized to modify financial or other commercial terms unilaterally? What actions require review and approval by the other party? Who can commit the joint resources of the partners when and as needed to support the sales effort? Are there geographical restrictions on the marketing efforts of either party? Although these questions may appear obvious, they or similar issues have been the basis for significant difficulties in actual joint sales efforts.

Even with a very competent partner and limited sales resources of one's own, do not abdicate participation in the sales effort. The real advantages of a third party sales effort lie in the complementary strengths of the partners. An inactive partner may cherish the illusion of a "free ride," but, in reality, may be inadvertently trapped into some very costly commitments because of this passive role in the proceedings. These difficult situations usually arise, not because of any evil intent on the part of the active partner, but rather of the absence of the perspective and contributions of the passive partner.

Finally, and perhaps obviously, set up a reliable and mutually understood communications system between the partners. Do not depend on the judgment of each partner in the effort to determine the need for a communication or "bulletin" of some sort. Remember that the sense of what is significant and important may differ greatly because of the previously-mentioned difference in motivation. Establish a mutually acceptable system of periodic reports to ensure a common understanding of current activities. If the reports serve only to prevent a few unpleasant surprises, they will have justified the effort.

9

The Licensing Agreement

In general, it can be said that all the activities discussed thus far are directed toward the subject of this chapter, The Licensing Agreement. Market research, product pricing, packaging, sales--all these, and other activities as well, are directed toward a licensing agreement, which is the instrument that governs the actual technology transfer. Having said that, however, it becomes abundantly clear that there is no such thing as a "typical" license agreement, because of the wide variety of transactions involved in technology transfer.

Consider first some of the variations in the broad range of technology transfer activities. There have been, for example, true technology transfers that involved only the delivery of a notebook of data and instructions, and perhaps a brief presentation reviewing the documents. In spite of the abbreviated nature of the transaction, the transfer was carried out under the terms of an executed licensing agreement, covering such subjects as rights of the parties, fees and other financial considerations, and continuing obligations, such as secrecy.

At the other extreme would be the more complex transactions, involving grants of freedom to practice under certain patents, complex schedules of payments and royalties, rigorous guarantee and performance tests, provision for future transmittals of technological improvements by both licensor and licensee, as well as a broad range of technical and support services during start-up and commission.

Compounding the quest a "typical" licensing agreement is the potentially large number of contractual interfaces that may be associated with technology transfer. The key agreement, of source, is the technology transfer document itself. Associated with it, however, and affecting some of its provisions are such companion contractual arrangements as engineering services, project financing, product take, and raw material supply. An extreme example of these multiple contractual arrangements involves one petrochemical facility in south Texas wherein there were contract documents covering technology transfer, design, construction services, feedstock supply, energy supply, and waste disposal services. The key point, of course, is that licensing agreements, in addition to covering a broad range of technology transfer arrangements, may also have to mesh with other related agreements and thus can vary considerably in scope and content.

In spite of these variations, there are indeed certain elements that are common to many technology transfer agreements. An understanding of these elements or sections can serve as a very useful checklist during the marketing phase of the effort to ensure a consistent focus on the ultimate objective, that of an effective and profitable technology transfer.

For the purpose of this discussion, these common elements or agreement sections are as follows

   o   The Grant

   o   Supporting Services

   o   Confidentiality

   o   Payments and Fees

   o   Tax Considerations

   o   Patent Issues

   o   Guarantees

o Improvements and Grantbacks

o Records

o Assignment

o Termination

o Appendix Material

The intent here is not in any way directed toward a "do-it-yourself" approach to contract drafting. As with any contractual arrangement, the preparation of a licensing agreement should be left to your attorney. Nevertheless, it is essential that anyone with a key role in technology transfer have a clear understanding from the outset of licensing efforts as to the issues that must be addressed in a license agreement. An understanding, for example, of the Support Services and the Guarantee implications can provide valuable guidance in projecting technical resource requirements for the effort. By the same token, an understanding of the alternative approaches as to what is being delivered under the Grant section of the agreement can be a very useful negotiating tool.

Let us now take a closer look at these specific sections of a licensing agreement, keeping in mind that these are, by no means, all-inclusive or even common to every agreement.

THE GRANT

This is the section of a license agreement that addresses the very significant question of what exactly is being delivered by the licensor to the licensee in return for some rather substantial payments and fees. A clear and common understanding of this issue is the crux of the whole technology transfer effort, inasmuch as the type of grant can vary, depending on the technology and the needs of the parties involved.

In general, this section involves a grant by the licensor to the licensee of the right to practice his (the licensor's) technology, usually in connection with the manufacture, use, and/or sale of some desired product. If there are patents, the Grant section will provide for immunity for the licensee for infringement of the licensor's patent rights. In contrast to Grants extended in other commercial transactions which involve a commitment to *do something*, the principal grant in a license agreement entails a commitment *not to do something*, namely, bring suit for patent infringement. In spite of this negative connotation, the grant of freedom-to-operate, so to speak, does amount to a very valuable consideration for the licensee. Implied in the grant, and usually spelled out early in the agreement, is the contention that the licensor does indeed have clear title to the technological asset in question, and thus has the right to make the grant of freedom to practice.

A straightforward grant of freedom to operate under this licensor's patent rights may not be sufficient for the commercial needs of the licensee. In addition, there may arise during negotiations a need to enhance the perceived value of the technology being transferred. Because of these considerations, the language in the Grant section is sometimes modified to make the grant *exclusive*. This exclusivity provision is usually extended with certain constraints, such as limiting the grant of exclusivity to a particular geographical area. Exclusivity obviously involves trade-offs by the licensor of future business for more immediate prospects, and clearly illustrates the value of an understanding of market growth and future economic prospects in the area in question.

A grant of exclusivity may affect the financial and tax treatment of the royalty payments received from the transaction. If the grant of exclusivity does not include provision for retention of a royalty-free license by the licensor (i.e., all the licensor's rights are relinquished in a specific geographical territory), the proceeds from the transaction may be considered as capital gains from the sale of an asset rather than ordinary income, and are treated accordingly for tax purposes. Depending on the applicable tax law, and future licensing prospects in the territory in question, a grant of exclusivity can be not only a bargaining device, but may provide a significant financial advantage for the licensor. In these cases, a grant of exclusivity can thus be an attractive feature for both parties to the transaction.

A second enhancement of the value of the Grant is the addition of the right to sublicense. Here the considerations associated with the added grant are considerably more complex. For the licensor, who will presumably share in the sublicensing income, it offers a method of extending his marketing reach. On the other hand, it may involve additional training and support, as well as the potential for extended liability. In addition, a sublicensing arrangement can substantially complicate both technical and commercial interfaces. For the licensee, the right to sublicense can provide what amounts to an "equity position" in the technology, and thus a strong incentive to promote the product. A major consideration in any event, of course, is the technical and commercial capability of the licensee, and thus his future impact on the marketing effort for the technology.

Note that in most cases the licensor is not "selling" his technology. As the name of the section implies, he is merely granting the right to use the product and is not changing his status as owner of the technological asset. An actual sale of the technology, of source, would create a completely new ownership situation and would be handled as a sale of assets rather than as a straight licensing agreement.

SUPPORTING SERVICES

A licensing agreement must deliver at least two critical items in order to accomplish the technology transfer. One, the right to practice the technology, is essential, and has been outlined in the preceding section on Grants. However, the right to practice is useful and valuable only if there is also a transfer of knowledge or know-how relating to the technology in question. This know-how can be transferred in a number of forms: design packages, training sessions, models, review comments, operating manuals, analytical techniques, and a host of other devices. Although differing significantly in form and content, they share a common objective of transferring the technology to the licensee in a usable form and, for the purpose of this discussion, are referred to as Supporting Services.

The level of supporting services required is a function of both the nature of the technology and the needs of the licensee. A technically sophisticated licensee may require only a basic outline of the technology and its application in order to reduce it to practice. A Japanese licensee of a petrochemical process,for example, took the basic design packages and proceeded to design, commission, and successfully start up a commercial manufacturing facility with no further input or assistance from the licensor. Others, on the other hand, may require extensive training and detailed step-by-step procedures to make the transfer work. The important point, however, aside from the level of support required, is that the question of supporting services be addressed early enough and extensively enough in the technical and commercial discussions that a mutually satisfactory program can be defined and included in the agreement.

All of the costs of the supporting services will not usually be covered by the license fees or royalties. A process package, such as that described in the chapter on product packaging, will be offered sometimes for a supplemental fixed fee, or sometimes on a time and materials basis. Commissioning and start-up help, in many cases, will be provided on a per diem basis, over and above a fixed minimum amount. In many cases, the financial arrangements and other contractual considerations related to these supplemental services will be spelled out in an appendix section of the licensing agreement.

It cannot be emphasized too strongly that the proper selection and delivery of supporting services are crucial to the success of the technology transfer effort. All too often, the licensee will perceive the supporting services as an attempt on the part of the licensor to extract more money from the transaction. Because of this, he may skimp on needed elements of the technology transfer, sometimes with very undesirable consequences. The "do-it-yourself" syndrome, sometimes initiated by the licensee as an economy measure, has been the culprit in a number of technology transfer problems. By the same token, there have been instances in which the licensor has tried to minimize the supporting services he offered in order to preserve scarce technical resources for other uses. For example, the same technology that, as noted above, was successfully licensed in Japan on a minimum support basis, was subsequently licensed to a considerably less sophisticated licensee, and the licensor minimized the supporting services offered because of a serious shortage of technical resources. The result was an incredibly difficult start-up, together with severe cost penalties for all involved. It has been stated before, but bears repeating, that a licensing agreement with halfhearted or inadequate technological support can be a ticking technological time bomb, with commercial consequences that far exceed any cost savings foreseen. Because of these considerations, some licensing agreements include a certain minimum level of supporting services as a mandatory part of the technology transfer, as a means of ensuring effective start-up and performance in accordance with expectations.

CONFIDENTIALITY

The value of technical information and know-how is clearly related to its restricted availability. If the information is generally available to all, its perceived value is jeopardized, to say the least, and the potential for imitation or patent claim avoidance is probably greater. It is, of course, true that patent law requires disclosure as a prerequisite to the statutory protection offered under the patent system, and that one "skilled in the art" should be able to practice the technology from the disclosure in the patent. Nevertheless, the likelihood of a successful commercial design of a manufacturing facility based solely on a patent disclosure is rather slim, inasmuch as it is usually impractical to include all associated know-how and experience in a patent disclosure. For this reason, confidentiality and secrecy considerations are important elements in a licensing agreement.

In many instances, the licensee will have already executed a secrecy agreement in connection with previous preliminary disclosure made for evaluation purposes. However, because very significant amounts of additional information will be made available under the license agreement, a separate confidentiality clause is included in most licensing agreements.

These clauses usually outline the intention of the licensor to disclose valuable technical information to the licensee for the purpose of practicing the technology. In some cases, a broad description of technical information is included in the definitions section of the agreement. The confidentiality clause will usually restrict both the disclosure and the use of the information thus transferred. Certain material will be specifically expected from the restrictions. These exceptions usually include material already known to the recipient without secrecy constraint, material subsequently becoming public knowledge through no fault of the recipient, and material received without secrecy agreement from those having a bona fide right to disclose.

Provisions are usually included in the confidentiality clause that permit the licensee to disclose the confidential information, as needed, to contractors or consultants who may require the information to carry out assignments for the licensee. These provisions usually include a requirement that the contractor or consultant be bound to secrecy to at least the same extent as the licensee himself.

In most confidentiality agreements, the obligations of secrecy fall specifically on the licensee as the party executing the agreement, with the implication that the employees of the licensee will be themselves bound through pre-existing obligations to their employer. However, in some cases, particularly where there is concern regarding job mobility and employee turnover, individual secrecy agreements may be required of those employees involved in the evaluation or technology transfer.

Most confidentiality clauses in licensing agreements, as well as freestanding secrecy agreements, will include a termination date for the secrecy obligations. Because we have here a very quantitative expression (term or termination date) of a very qualitative judgment (what is a reasonable period of secrecy?), this element of the confidentiality clause is often the subject of considerable debate. Most licensors will argue vigorously that a long period is necessary to preserve the value of the particular technological asset in question. The contention of the licensee, on the other hand, is that long periods of secrecy are unrealistic in view of natural attrition and the rapidly changing nature of most technologies. Although the secrecy terms can vary considerably, a period of 5 to 10 years is not unusual in most agreements.

In connection with the duration of secrecy constraints, either in the licensing agreement or in a stand-alone secrecy agreement, care should be taken that these time periods are

not confused with other periods or termination dates that
may be referred to in the same or related agreements. Most
agreements, for example have provision for termination of
the agreement in the event of a failure to perform on the
part of one of the parties. In addition, there is usually a
finite period over which the licensee is obligated to pay
royalties. In connection with the definition of these various
time periods, there is usually a provision that the secrecy
obligations incurred by the licensee in the section on
Confidentiality will survive the termination of the agreement
for whatever cause, thus preserving the value of the asset.

All of the secrecy provisions in a licensing agreement
are not solely for the benefit and protection of the licensor.
In many cases with continuing support obligations, the
licensee may need to disclose technical and commercial data
to the licensor in the course of the technology transfer
activity. For example, the licensee's proprietary control
systems or analytical techniques may be exposed while the
licensor is reviewing the plant design. For this reason, there
is often a two-way secrecy obligation outlined in the
Confidentiality section, thus protecting the know-how of both
licensor and licensee.

Finally, there are often special restrictions on the
transfer of technical information that relate to the U.S.
Department of Commerce regulations against flow of critical
technology to certain foreign governments. In general, these
restrictions involve a statement that the agreement in its
entirety must conform to current regulations in this area,
including provision for an export license, if applicable.
There is usually additional language that prohibits the
licensee from sublicensing or engaging in other forms of
technology transfer that may involve prohibited recipients.
In this connection, it is important to note that an
understanding or, at the very least, an awareness of
technology export procedures and restrictions of the
Department of Commerce and other governmental bodies can
be quite important in even the early stages of the sales
effort. These regulations can apply well before any license
agreement is executed or even drafted, inasmuch as
disclosures of certain types of information can be constrained
by these regulations. See the Appendix of this book, or
check with the Commerce Department, so that the need for
further action can be defined and established early in the
sales effort.

PAYMENTS AND FEES

In the first two sections of this chapter on licensing agreements, we have discussed Grants and Supporting Services, the benefits being delivered by the licensor to the licensee. In return for these benefits, the licensee obligates himself to some form of payment, usually, but not always, in the form of financial compensation. Because technology transfer transactions usually involve large fees and payments and obligations that can cover a span of many years, this section of a licensing agreement is a critical one.

The key elements in this section are, of course, the sums of money to be paid by the licensee to the licensor in return for the benefits discussed above. As we discussed in the chapter on product pricing, a combination of a lump-sum payment or payments, followed by running royalty payments based on the number of pounds produced, is probably the most common form. The lump sum, or "up-front" money, is usually payable in several installments due at various checkpoint events, such as execution of the license agreement, at fixed intervals thereafter, and mechanical completion. The amounts of the payments are defined, place of payment is specified, and, in many cases, grace periods and interest or penalty payments for late remittances are outlined. The currency in which the payments are to be made is also stated in the agreement. Because in some cases the period between contract execution and, say, mechanical completion can be long and difficult to predict, provision is sometimes made for escalation of the future payments in accordance with some commonly understood index reflecting inflation. Depending on previous transactions and agreements, credit may be granted against these "up-front" payments to the extent of previous payments for options or disclosures.

The running royalty payments are usually expressed as a percentage of the net sales price per unit of product, even though they may have been based on a share of the profit margin per unit. Care should be taken to define carefully the meaning of net sales price, i.e., how allowances, discounts, and other variations are to be factored into the calculation of royalty payments. By basing the running royalties on a percent of sales prices, instead of a fixed amount per pound, there is at least a presumed allowance for the effects of inflation and currency values over the years. Also usually included in this section is the period of time over which royalty payments are to be made. As in the case of the "up-front" payments, the timing of the royalty payment (usually quarterly) is specified, as well as grace periods and penalties. Finally, to protect the licensor somewhat against unforeseen economic events, some agreements will specify a minimum quarterly royalty payment that applies irrespective of the actual units of production. This provision can be the subject of considerable debate, but is usually justified on the grounds that the future running royalties are actually deferred payments for services and benefits previously delivered.

Where the form of payment is related to the units of production over a period of several years, as in the case of running royalties, provision is usually included for access to records during normal business hours, reporting, and any auditing procedures required to verify payments due for a particular time period. Associated with this is a clear indication of the form in which payments are to be made and the office to which payments and associated reports should be directed.

Note that in certain instances, such as when dealing through a third-party engineering contractor in some developing countries, the payments spelled out in the agreement with the licensee may be for the cost of the total plant. These payments will include, but not necessarily identify, the fees and payments associated with the technology itself. These technology fees then will themselves be the subject of a separate license agreement between the licensor and the engineering contractor, which will include a restricted right to sublicense.

Thus far, we have limited our discussion of payments and fees to those in monetary form. There are, of course, technology transfer transactions wherein payment is made in some other form. For example, in technology transfer associated with a joint business venture,the partner supplying the technology may be paid with an equity position, such as common stock in the venture. In cross-licensing or similar transactions, the payment may be in the form of rights to other similar or complementary technology. Also, as discussed in Chapter 7 on product pricing, payments for some technologies may be accomplished by taking product from the facility employing the technology.

Finally, related to the general area of payments and fees, although sometimes discussed elsewhere in the agreement, is the "most favored licensee" clause. In essence, this assures the licensee that he will receive the benefit of any more favorable commercial terms extended to any future licensee. For future transactions, this clause can provide a "floor," or practical support level, for the payments and fees from future recipients of the right to practice the technology.

TAX CONSIDERATIONS

An in-depth treatment of the subject of taxation of intellectual property transactions in the various countries of the world is well beyond the scope of this book. However, there are some recurring tax-related issues that anyone involved in the technology transfer should consider carefully in the discussions and negotiations leading to a licensing agreement. Although the language addressing these issues in the agreement will be drafted by an attorney, all involved in the transfer effort should have some understanding of the effect of tax considerations on the structure of the agreement. For this reason, we shall proceed to take a brief look at the following tax-related issues:

o   Tax withholding in international transactions

o   Capital gains versus ordinary income

o   Deferred payments

o   Differing tax treatments for licensing elements

Tax Withholding

In many countries, a portion of technology transfer payments to foreign licensors will be withheld at the source to cover tax liability for income earned in that area. Up to a certain point, these withheld taxes can be used to offset tax liabilities in the licensor's home country, so that there is no net additional tax liability incurred. However, it is possible that the level of the withholding rate or other factors will preclude a complete offset, and the payments will thus be subjected to what amounts to double taxation. In addition, future tax changes in the licensee's country can lead to completely unanticipated additional taxes that can significantly affect the profitability of the total venture.

For this reason, language is sometimes included in the agreement that recognizes the existence of withholding on royalties or other payments, but shifts the liability for excess withholding (that which cannot be offset against the licensor's domestic tax liability) to the licensee. Obviously, the final resolution of this issue is subject to negotiation, but there should be a clear understanding of the potential consequences of withholding by the licensee's government on the total profitability of the licensing venture.

Capital Gains Versus Ordinary Income

In most technology transfer transactions, all property rights of the owner/licensor are not conveyed to the licensee, just as a landlord retains rights in his rental property in spite of its use by the tenant. In both cases, the income is derived from the use of a retained asset, and is recognized, for tax purposes, as ordinary income. (It is recognized that the tax treatment of capital gains will differ from time to time and place to place, so that tax differences, if any, between capital gains and ordinary income may vary considerably throughout the world.)

However, if the licensor relinquishes all rights in the technological asset as, for example, in an outright sale of the technology, the income received, after modification to reflect the cost basis, can be treated as a capital gain. An outright sale, though, is not always required to qualify the income for capital gains treatment. For example, a license with geographical exclusivity for a particular country, and in which the licensor retains no residual rights, can qualify the income derived as a capital gain. Great care should be exercised that restrictions, such as a limited field of use for the technology, are not placed on the rights thus conveyed, or the revenues may be treated as ordinary income. Keep in mind also that payments for services, such as start-up assistance, will probably be treated as ordinary income, even though the payments for the transfer of the technology itself may qualify as capital gains.

Deferred Payments

Care should also be exercised in scheduling deferred payments in technology transfers that qualify for capital gains treatment. If, indeed, deferred or installment payments are built into the agreement, a portion of these payments may be classified as "imputed" interest and thus taxed as ordinary income. The presumption is, of course, that the deferral of payments is a significant consideration reflecting the time value of money.

The imputed interest consideration is applicable to any payments made more than six months after the sale or license agreement. The amount of imputed interest is fixed by statute, and varies depending on the time period in which the transaction occurs. For transactions in the early 1980s, for example, interest would be imputed at a rate of 10 percent, compounded semiannually. Thus, in some instances, the tax treatment of imputed interest can be a factor in structuring deferred payment terms.

Differing Tax Treatments For Licensing Activities

When developing technology transfer agreements with licensees in certain areas of the world, such as certain Latin American countries, care should be taken to differentiate payments for the technology itself from certain payments for specific services. For example, in Argentina a portion of the payments for technical assistance or engineering services is exempt from taxation if the services cannot be obtained within the country. The effective tax rate for these items is thus lower than that applicable to royalty or lump-sum payments for the right to practice the technology, and it is therefore beneficial, from a tax standpoint, to identify and price such services separately in the licensing agreement. In general, most feasibility and engineering studies, as well as commissioning and start-up assistance, are eligible for the reduced tax rates in those countries that differentiate on tax treatment.

PATENT ISSUES

As we discussed previously in this chapter, one of the principal benefits received by a licensee under a technology transfer agreement is the right to practice technology protected under certain patents held by the licensor. The Grants section of the agreement spells out this freedom to practice in very specific terms. However, this freedom to practice is extended only by the licensor and can be affected in various ways by the actions of third parties who are not bound or affected by the agreement. Although there are many patent-related issues, two of the principal issues that should be addressed in the agreement are as follows:

o   Infringement by others

o   Patent litigation by others

In this section, we shall discuss some of the agreement provisions that one should consider in relation to these two issues, as well as some other patent-related issues that may be pertinent to the technology transfer effort.

Infringement By Others

Regardless of how extensive the licensor's patent coverage for the technology may be, there always exists the possibility of a third party practicing some aspect of the technology without the benefit of a license to do so. Inasmuch as an infringement of this nature significantly affects the value of the benefits received by the licensee under the agreement, the responsibilities and actions required by licensor and licensee with regard to third-party infringement are usually addressed in the license agreement.

In this section, the agreement will usually define who (usually the licensor) initiates and manages the infringement proceeding against the third party, as well as who bears the cost of the effort. In addition, it is usually desirable to specify if one party has a unilateral right to settle the action, and how any payments recovered from the third party as a result of the litigation are to be apportioned between licensor and licensee(s). Another significant question to be addressed is the impact of the failure of the infringement litigation on the licensee's continuing obligations under the agreement. The answer to these questions, of course, is a function of the value of patent protection relative to other benefits received under the license agreement, the potential financial impact of infringement, and, in many cases, the negotiating skills of the parties to the agreement.

Patent Litigation By Others

The infringement question, of course, can work both ways. It sometimes happens that the licensee is confronted with an infringement action by a third party claiming to hold patent right in the technological area in question. It is important that this or related issues be addressed in the license agreement, even if the potential for such a problem appears to be unlikely.

One response to this problem is to include in the licensing agreement a provision whereby, upon written notice by the licensee, the licensor will undertake at its own expense the defense of infringement actions brought by third parties. Along with this provision, the licensee is usually required to provide assistance to the defense effort. In some cases, the licensor will agree to pay some portion of any judgment assessed by the court in the infringement proceedings, usually up to some percentage of the total license fees. The ultimate problem arises, of course, if the licensee is required to discontinue the practice of the technology, and the licensor understandably attempts to limit the liability for this eventuality.

Another patent issue, related to but not included in the major categories discussed above, relates to an admission on the part of the licensee as to the validity of the patents cited in the license agreement. This provision is intended to protect the licensor from future actions by the licensee to contest the validity of the applicable patents, and thus eliminate or reduce his obligations under the license agreement.

GUARANTEES

The subject of guarantees is sometimes likened to religion and politics from the standpoint of its emotional impact and the widely differing attitudes and value judgments prevailing throughout the world. In the case of most sophisticated licensees, there is a clear recognition that guarantees seldom, if ever, provide adequate protection from a catastrophic failure to perform by a technology product. The cost of the lost revenue because of business interruption related to faulty technology, not to mention possible facility revamps, will far exceed any guarantee coverage extended by most technology transfer agreements. These guarantee penalties are usually limited to some portion of the total license fees and royalties or, in some cases, the cost of design work (not material and labor) required to effect remedial measures. In many instances, the major impact of a guarantee provision is to provide performance motivation for the licensor. Even here, however, the motivation arising from guarantee obligations may be far less than that related to professional reputation and its impact on future business prospects.

Nevertheless, guarantee provisions as part of a licensing agreement do serve a useful function in providing a reference point and a means of "keeping score" regarding effective transfer of the technology. The guarantees quantify, in effect, the performance goals of the technology and thus define the benefits to be delivered to the licensee. In some cases, proposed guarantee levels for such variables as raw material and utility consumptions are used as a basis for technology comparison and selection. However, there should be a clear understanding as to whether the guarantees are based on past commercial performance or on a "hard-sell" risk-taking posture by the would-be licensor.

The performance factors on which guarantees are provided will depend on the technology being transferred and the economic impact of the factors on the success of the venture. Probably the most common items subject to guarantees are production capacity, raw material and energy usage, and product quality, depending to some extent on the amount of control the licensor has over facility design.

Guarantees for raw material consumption, for example, will usually be stated in terms of units of feedstock consumed per unit of refined product produced. Conditions under which the consumption is measured however, must be carefully defined. First, conditions for a performance guarantee run must be specified in order to ensure that data are obtained under representative, steady-state conditions. Second, it is essential that methods of measurement be defined so that there will be a common understanding of the basis for the consumption and production quantities used for the performance determination. Third, involvement and access by the licensor's representative during the performance run should be specified. Finally, there should be a finite time period after start-up during which a performance guarantee test run must be carried out.

Similar qualifications are desirable for the other performance factors subject to guarantees. Great care should be taken to define measurement methods, particularly analytical techniques. Also, one should be alert for performance factors, such as energy consumption, that can change after agreement execution during the course of design trade-offs. For example, a heat exchanger might be eliminated as a capital cost reduction measure, thus increasing steam or fuel gas consumption above the original design basis. Guarantee provisions should thus be drafted with specific reference to the original design basis, or with sufficient flexibility to accommodate future design changes and trade-offs.

Along with the guarantee provisions, the licensing agreement will provide for financial penalties for failing to meet performance targets. For example, every unit of raw material consumption in excess of the guaranteed amount may entail a financial penalty amounting to a specified percentage of the license fee. Similar penalties would be provided for other guarantee components that fall short during the performance test run. In most cases, there is a contractual limit on the total amount of the guarantee penalties, usually amounting to one-half or two-thirds of the total fee.

In developing the guarantee section of a licensing

agreement, care should be taken to ensure that credit is received for performance in excess of that set forth in the guarantees. This is usually accomplished by providing for offsets, wherein, for example, energy consumption in excess of that guaranteed can be offset to some extent by raw material consumption below that which had been guaranteed. The offset relationships (i.e., amount of raw material consumption offsetting a specific shortfall in energy consumption) can be rather tedious to develop, but can serve to alleviate significantly the potential guarantee penalties associated with a particular technology transfer effort.

IMPROVEMENTS AND GRANTBACKS

Unlike most tangible conventional products, technology products or intellectual property can evolve or grow into more valuable assets as the result of improvements realized through commercial practice and continued R&D efforts. Moreover, there is a strong incentive for both licensor and licensee to allocate resources for the purpose of improving and strengthening the technological product. For the licensee, such improvements will obviously be reflected in the economic performance of the venture using the technology. For the licensor, technological improvements can serve to enhance the competitiveness and value of the licensed technology product and to ensure a continuing stream of revenue from the product. Because of these considerations, most licensing agreements include provisions for the encouragement and management of improvements to the technology. These provisions are usually categorized under the general heading of improvements and grantbacks.

Grantbacks apply to improvements made by the licensee to the licensed technology and, as the name implies, involve the "granting back" of a royalty-free, non-exclusive license to the licensor for the practice of the improvement in connection with the use of the original technology. The grantback usually includes the right by the licensor to disclose and sublicense the improvements to other licensees of the same technology. Although this may at first appear to negate any competitive advantage achievable by the licensee from his technological efforts, it does provide him with access to future improvements developed by the licensor or other licensees of the technology. Some licensees will choose to pursue their own improvement programs and to forego, if possible, the grantback arrangement. Still others will welcome such an arrangement as providing convenient access to an otherwise unavailable pool of technological resources.

Closely related to the grantback provisions is an arrangement whereby improvements made by the licensor after the execution of the agreement are provided to the licensee without the payment of additional fees or royalties. Sometimes referred to as a "grant-forward" provision, this arrangement assures the licensee that he will have access to all improvements and innovations provided to subsequent licensees during the term of the agreement, and is, in a sense, the quid pro quo for the grantback provision discussed above.

One effective grantback and "grant-forward" mechanism often used with multiple licensees is that of the periodic licensee meeting. These meetings are usually organized around the presentation of a series of papers by both licensor and licensees on various facets of the practice of the technology. The papers cover not only recent advances and improvements to the technology, but also address areas of mutual interest, such as materials of construction, safety, and waste disposal. In order for the meetings to be effective, a great deal of preparation and management effort is required, and there are sometimes extensive discussions regarding the distinction between proprietary data and data falling under the grantback provisions. In spite of these differences and the extensive amount of management effort required, most participants feel that the information gained and the exchange of views more than justify the resource requirement.

Finally, provision should be made in the agreement for patent coverage on the technology improvements thus developed. In most cases, patent coverage is pursued by the licensee developing the improvement. However, in some instances, that party may not feel that the effort and expense of obtaining a patent are warranted, and the agreement may give the licensor the option of pursuing the patent.

RECORDS

Where running royalties are to be paid over a period of years, recordkeeping becomes something more than a pure clerical concern. There should be a clear understanding of what reporting is required of the licensee and at what intervals the reports are to be made. In addition, access to the records of the licensee for inspection and auditing should be clearly defined in the agreement. Agreement on the broad concept of the basis for fees and royalties is not usually sufficient unless there is a clear definition of the implementation mechanics, including recordkeeping. Some careful effort and attention to the detail of accounting and reporting procedures in the contract development stage can yield significant benefits in future agreement administration.

ASSIGNMENT

Most agreements will include language that restricts the right of the licensee to assign the rights and privileges granted under the license. In most cases, these rights are extended to successors to the parties to the agreement, and thus changes in ownership of the business involved in the practice of the technology can be accommodated.

TERMINATION

We have thus far considered a number of sections in the licensing agreement wherein the obligations and benefits of the contracting parties are defined. These include areas such as the right to practice the technology, payments and fees, access to improvements, and secrecy, among others. The primary emphasis, understandably, is on the description of the benefits and obligations, as well as the starting dates, i.e., when information becomes available, when payments are due, etc. However, it is equally important that there be a clear understanding on the part of both parties to the agreement as to when these benefits and obligations terminate.

If the license agreement entails the right to practice

technology covered under various patents, that right may terminate with the expiration of the patents. At the same time, however, there may no longer be a need for a special right to practice in the absence of additional patent coverage.

Royalty payments usually continue over a period of years, and the termination of this financial obligation must be spelled out in the agreement. Accordingly, the agreement will usually specify that, upon fulfillment of the royalty payment requirement, the licensee will have a paid-up license to practice the technology within the restrictions (plant size, geography, etc.) set forth in the grant.

Secrecy obligations do not usually expire concurrently with the other obligations and benefits defined in the agreement. In fact, many agreements will contain specific language noting that the secrecy commitments will survive the expiration of other terms of the agreement.

Finally, most agreements will contain a provision for early termination of the agreement in the event of default by one of the parties in the performance of contractual obligations, such as prompt remittance of royalty payments. This section usually provides for written notice and an opportunity for remedial action prior to termination. It is important to specify in this section those obligations, secrecy, financial, or others, that will survive the termination.

APPENDIX MATERIAL

Because they are related, but not necessarily pertinent to the main body of the technology transfer agreement itself, certain supplementary materials are sometimes included as attachments, or as an appendix, to the license agreement itself. Some of the types of material thus included are technical service and support agreements, definitions of technology packages to be provided, details of performance guarantees, and analytical and measurement techniques required to define performance under the agreement.

10

Licensing Support Activities

All too often, it may appear that the licensing effort is essentially complete with the execution of the license agreement. Certainly, a number of critical activities have been successfully completed, a marketing concept has been transformed into a commercial reality with prospects for significant streams of income over future years, and the yield on an intangible asset has been significantly increased. Given these favorable events, then, it would not be too surprising if there were a reduced emphasis on the supporting activities required after the license agreement has been executed.

In reality, however, the numerous operational and administrative tasks remaining after the execution of the agreement probably comprise, more than any other activity discussed thus far, the keys to success or failure of the entire technology transfer effort. A halfhearted or less than completely professional licensing support effort, at best, will inhibit future commercial prospects and may also create serious legal and financial problems for both parties. A carefully prepared and staffed licensing support program, on the other hand, can ensure that both parties emerge from the technology transfer transaction as "winners" with their commercial expectations realized.

Organizationally, licensing support activities are handled in a wide variety of forms and structures. The ultimate, in a sense, is that used for one-product, multi-licensee, continuing long-range programs wherein there is a permanent staff of technical specialists and others to provide specialized, experienced licensing support. The organization is a permanent, ongoing one with a clearly defined mission and with career development and growth opportunities for its members.

Perhaps a more typical situation, however, is that which prevails where there may be a multiplicity of technology products, and possibly as few as one licensee per product. Here one does not have the opportunity of staffing with a cadre of "experts," but must compete instead with other organizational requirements for the critical resources required for licensing support.

The point of all this is not so much the "right" organizational form as it is the need for recognizing the highly critical role occupied by licensing support activities in the overall technology transfer strategy. Whatever the organization, there must be a clear understanding from the earliest stages of a licensing effort that these support resources will be required, and a plan devised to acquire them. A recognition of this need only after the license agreement is executed will most assuredly detract from the effectiveness of the technology transfer effort, and can also lead to serious difficulties with resource allocation within the organization itself.

As previously noted, there is a broad spectrum of licensing support activities required for an effective technology transfer program, although the full range may not be required for every transaction. The depth and breadth of support required will be determined by a number of variables, not the least of which are the technological sophistication of the licensee and the complexity of the technological product. In general, however, licensing support activities fall into two major categories: operational support and administrative support. More specifically, operational support is that directly applicable to a particular technology transfer transaction, and is usually spelled out in the license agreement. Administrative support, on the other hand, is related to the ongoing management effort directed toward continuity and effective response in the technology transfer area. Let us now take a closer look at these two important support areas.

OPERATIONAL SUPPORT

Operational support, involving such activities as the preparation of process packages, review of detailed design drawings and specifications, and assistance in the commissioning and start-up phase, is one of those activities for which there is a considerable body of intellectual support. There is, for example, general agreement that prompt and effective preparation of a process package, such as that outlined in Chapter 5, Product Packaging, is an essential element in most technology transfer transactions. The reaction, however, may be considerably different when an unexpected request for operational support, such as process package preparation, is presented to the director of engineering to be worked into an already overcrowded agenda. As a general rule, early involvement of functional managements such as those of engineering, R&D, and manufacturing, in the licensing effort is essential to effective operational support. All too often, because of uncertainty regarding the outcome of the technology transfer efforts or, perhaps, benign neglect, the licensing manager will fail to keep functional support management involved or even informed in the early stages of a licensing activity. The result of this failure to communicate can have a serious negative impact on the kind and level of operational support provided. If there is one clear message related to operational support, it is the continuing need for deliberate early involvement and communications with those functional managements who will be needed to support the technology transfer effort. The message may appear to be obvious in this age of instant communication, but past case histories demonstrate a chronic problem in this area.

There are also some other general guidelines applicable to the area of operational support. Some recent interviews with experienced licensing technical support personnel have highlighted several critical areas. Although not necessarily applicable to all situations, they provide a useful proven checklist for future technology transfer transactions.

o   Try to ensure that technical support
      personnel have at least a general
      understanding of the commercial terms
      and provisions. In many instances, they
      will be situated such that they will not
      have access to commercial guidance and
      interpretation from their own
      organization.

o   So-called commercial discussions nearly
      always have technological overtones.
      Conditions for guarantee runs, raw
      material supply arrangements, sourcing
      for equipment supply—these issues and
      others make it desirable to involve
      technical personnel in an early stage of
      the discussions.

o   Continuity, like motherhood, is an almost
      universally acclaimed institution, and
      this is particularly true in operational
      support. The availability of technical
      personnel with background and
      experience in technology transfer can
      enhance the probability of success in
      even the most difficult transaction. As
      is often the case, this is a classic
      "easier said than done" situation.
      Nevertheless, there are some approaches
      that can help. Develop and maintain a
      file of knowledgeable personnel in each
      technology area so that requests for
      support can be specific. Feed back to
      the individuals' management their
      contributions and achievements in the
      technology transfer area. Use special
      assignments, where practical and cost
      effective, to extend and enhance the
      individuals' relationship with the
      licensing function.

These general guidelines must be supplemented with a
careful consideration of the specific areas of operational
support. Although these may vary, the principal types of
activity are the following:

o   Process package preparation

o   Detailed design review

o   Commissioning and start-up assistance

The role of each of these and guidelines for implementation are outlined in the following sections.

Process Package Preparation

The process package is, in many instances, the principal instrument by which the technology is transferred. Reference is made to the section on transfer packages in Chapter 5 for a description of the content and format of a typical process package. Because it is a key element in the technology transfer process, a clear understanding of this activity is essential to effective operational support.

First, be sure there is a common understanding of exactly what the process package will or will not include. A *basic process package* or a *detailed process package* may not mean the same thing to all parties involved. Itemize, first for discussion purposes and then perhaps as an appendix to the license agreement, specifically what the process package will include. Ascertain, so far as possible, who will be using the package and what their needs are. Will the package be used primarily by the licensee, or will it instead be used by an engineering contractor as the basis for detailed design?

In addition to a clear understanding of content, establish at least a general outline of a basis for design. An understanding of battery limits delivery conditions for feedstocks and utilities, staffing and control systems philosophies, and laboratory and maintenance capability, to mention only a few, can go a long way toward ensuring a package that will transfer the technology effectively. Carefully define the applicable codes, standards, and fabrication practices that must be followed in the course of design. Be assured also that these efforts to prepare a compatible, workable package are not merely a philanthropic gesture; rather, they reflect an enlightened self-interest in ensuring a workable design and a plant that will produce the desired stream of royalty income.

Process packages can be part of the services included in the license fees. More often, however, they are priced separately and the terms spelled out as a supplement to the license agreement itself. Separate pricing permits flexibility in tailoring the package to fit the needs of a particular licensee. Pricing is usually expressed as a lump sum, but may be quoted alternatively on a time and materials basis. Although the price may include a nominal profit for the supplier, the major emphasis is usually directed toward ensuring an effective transfer of technology and thus protecting future licensing income prospects.

Detailed Design Review

The process package, be it an abbreviated one or a very detailed version, is transmitted to the licensee or his designated engineeering contractor for use as the basis for the detailed design of a manufacturing facility. Even with a very experienced contractor or licensee, it is usually not feasible to include in the process package all the information required to ensure absolute compliance with the technology requirements in the preparation of detailed mechanical, electrical, civil, and control systems drawings and specifications. For this reason, it is usually quite important that the detailed design documents be reviewed by a representative of the licensor who is familiar with the requirements of the technology.

Such a design review can be carried out in the office in which the detailed design is being performed, or, alternatively, the documents may be shipped to the licensor's offices and the review performed there. In most cases, this latter approach will result in a more comprehensive review, inasmuch as there will be an opportunity for review by a broader range of knowledgeable specialists. As a general rule, the primary review of the detailed design is by those who have developed the process packages, but review by selected other disciplines and specialists is usually desirable, particularly if the design departs somewhat from past practice.

Most license agreements (or supplemental appendix material) provide for the supply by the licensor of a certain number of man-days of review and monitoring effort, such as detailed design review, at no additional cost to the licensee.

Here again, the licensor's motivation is to ensure proper interpretation of the technology, and thus the success of the licensing effort. Additional review and monitoring time is usually made available on a per diem basis, in order to accommodate needs of individual licensees. Careful documentation of review comments and their disposition is a "must," as they may become an issue in future start-up and operational difficulties.

Commissioning and Start-Up Assistance

After having prepared a process package reflecting the technology, and having reviewed the detailed design to ensure compliance with the process requirements, the remaining critical area of operational support involves assistance with commissioning and start-up efforts after construction has been completed. Although not necessarily in the same time frame, this area would also include activities such as training the licensee's operators and providing necessary manuals, covering such subjects as maintenance, analytical procedures, and safety considerations.

Where applicable, operator training is usually conducted, at least for key personnel, by the licensor in his own manufacturing facilities. When such a facility is not available, classroom training is provided, sometimes in conjunction with hands-on training in some similar operating unit. Here, perhaps more than in other types of operational support, great care must be taken to tailor the training approach to fit the technological sophistication of the licensee. A clear understanding of the background and experience level of the licensee's work force will be essential to an effective start-up and commissioning effort.

In the training effort, sufficient technical information and safety considerations data are provided to serve as the basis for the preparation of the detailed, step-by-step operating procedures. However, the preparation of the standard operating procedures is usually carried out by the licensee's operating personnel as part of the training effort. In a similar vein, necessary checkouts of piping, instrumentation, and equipment installation by licensee personnel is made an integral part of the training program.

Where possible, training, commissioning, and start-up

support are provided by licensor personnel with manufacturing experience in the same or similar technology. Here again, early recognition and communication of the need for this critical resource to functional management is essential. In addition, where the plant site is in a very different and "exotic" part of the world, careful indoctrination and orientation of the trainers themselves can yield significant benefits in making the transition to a very different culture without costly turnovers and political blunders.

Commissioning and start-up assistance is usually covered in a separate or supplementary agreement. A certain minimum level of support is usually included as part of the technology transfer, with supplemental assistance on a daily rate plus expenses. Inasmuch as this support activity is usually at the licensee's site, considerable care should be taken to see that administrative considerations are properly covered in the agreement. Such issues as liability for personal injury at the other party's site, level of living—cost and travel expense coverage, availability of translators, language to be used in manuals, frequency of home visits, and so forth should be carefully considered and resolved in advance. The midnight shift during the early days of start-up is not the optimum time to be addressing these administrative issues.

ADMINISTRATIVE SUPPORT

Thus far, we have discussed licensing support activities in the context of specific licensing ventures. Undoubtedly, as we have mentioned before, this type of support is essential to the success and future well-being of the licensing program. In addition, it is not likely to be overlooked, inasmuch as there will be in all likelihood, an immediate response from the licensee in its absence.

There is also, however, another generic type of support that is needed to ensure a professional, well-managed licensing program. This is the co-called administrative support that serves to ensure contract compliance, organizational continuity, and an effective technology transfer interface for contacts with the outside world.

As in the case of operational licensing support,there is no one organizational pattern that defines administrative support and its functions. As a matter of fact, one of the principal criteria of effective administrative support is the ability to survive organizational changes and still provide continuity of the licensing program. When one considers the number of organizational changes likely to take place over the ten to twenty year term of a licensing agreement, the need for some continuity in administration is apparent. In view of these considerations, the clearest picture of administrative licensing support is obtained by a consideration of the activities included in this area.

Probably the key element in administrative licensing support is a concise, accessible data base containing abstracts of all licensing and related agreements, with particular emphasis on contractual deadlines and commitments. Although some license agreements are mercifully brief, others, for legitimate reasons, can comprise more than one hundred pages to set forth the provisions of the agreement. In many cases, during contract drafting and development, little consideration is given to the mechanics of accessing these voluminous documents over many years in the future to determine obligations, commitments, and deadlines required under the contract.

Computerized data base management programs are, of course, invaluable in addressing this problem. Whether computerized or not, however, the principle need is to define carefully the material to be abstracted from the original agreement. The following tabulation represents some of the key items for the abstract:

o   Technology licensed

o   Licensee

o   Effective date

o   Termination date

o   Grant, exclusivity, right to sublicense

o   Secrecy obligations and term

o   Other related agreements

o   Reporting obligations and timing

o   Payment basis and timing

o   Payment record

o   Actions taken and notice given

o   Patents involved and related data (may be cross-indexed to separate patent management file)

o   Audit rights and obligations

o   Profit center financial responsibility

The specific format and content, of course, must be tailored to fit particular organizational and management needs, but the above information is typical of most such data base systems.

Another requirement for an effective administrative support system is a knowledgeable, constant interface for external contacts and inquires. For small organizations, this is often the chief executive officer or his designee. For larger firms, however, the issue can be considerably more complicated, particularly if there are repeated reorganizations and product realignments. Attempts by an outside party to find a knowledgeable spokesman for a particular technology in some large corporations can be frustrating, to say the least. Of even more concern are actual instances of inadvertent violation of contractual commitments (such as exclusivity, for example) by new managements unfamiliar with past activities. Although usually detected before irrevocable damage occurs, there can be, at the very least, embarrassment, and in some cases significant costs.

In some instances, the response to this problem has been the establishment of a corporate technology transfer office. Although the office (or individual, in small organizations) would not necessarily have line responsibility for the decision to license or not, or for operational support, it would fulfill two important roles:

o   It would serve as a central clearing house for all information, obligations, deadlines, and action items related to technology transfer activities.

o   It would provide an ongoing external interface with the outside world for licensing inquiries and other technology transfer matters, irrespective of internal organizational changes.

The key element in administrative support of licensing activities is, of course, not the staffing or organizational structure, but rather the recognition of the need for continuity and active ongoing management of both past obligations and future inquiries and commercial contacts.

Part III

Technology Acquisition and In-Licensing

11

Technology Acquisition

No doubt a cliche, but nevertheless still true, is the old saying, "For every seller, there is a buyer." We have thus far concentrated on the selling, or licensing *out*, side of the technology transfer transaction. Many licensing organizations and individuals, however, find themselves involved just as heavily in licensing technology *into* their organization. Although we have dealt extensively with the benefits achievable by licensing technology out to others, the rewards of an effective, well-planned, and managed technology acquisition program can be even more significant. When one considers the critical role that acquired or licensed technology plays as the basis for multimillion dollar commercial ventures, it is abundantly clear that a thorough understanding of methods and approaches to technology acquisition, or *licensing in*, is crucial to success.

Even if one were not contemplating the immediate acquisition of technological products, in-licensing warrants attention because of the useful perspective it provides on the customer's needs and motivation when one is trying to market one's own technology. An understanding of a prospective buyer's point of view can be a powerful marketing tool when licensing out to others.

Because we have already discussed at some length many of the important aspects of technology transfer in previous chapters, and because, in many instances, these apply equally well to the acquisition of technology, our treatment of licensing-in will be somewhat abbreviated in the interest of avoiding repetition. Nevertheless, there are unique characteristics to technology acquisition, and we shall discuss these in subsequent chapters on the following:

o   Technology search

o   Technology evaluation

o   Managing the transfer

First however, let us examine some of the reasons organizations and individuals seek to acquire technology. There can be many incentives for the purchase of technology, and an understanding of these is a prerequisite to the formulation of search and evaluation techniques. In addition, an overview of these reasons may highlight some commercial opportunities that could otherwise be overlooked. Keep in mind that these reasons are not necessarily mutually exclusive, and that there may indeed be multiple reasons for the purchase of rights to technology from others.

SUPPORT FOR A NEW VENTURE

Probably the most frequent incentive for the acquisition of technology rights from an outside source is the need associated with a venture into a new product marketing area. An organization, for example, with a business strategy that includes entering the linear low-density polyethylene business, in all probability, will need to seek technology from an outside source. Similarly, many state-controlled economies, in order to implement their five-year plans for new manufacturing facilities, must allocate a significant effort to the acquisition of new technology from external licensors. In situations such as these, the decision to license in is almost a given, i.e., if one does not have the technology "in-house," one must look elsewhere. The resulting acquisition effort thus has the virtue of being focused on a defined technology area, with few alternatives.

INCREASED COMPETITIVENESS IN EXISTING OPERATIONS

The need for the acquisition of technology does not always arise from entry into new ventures. In some instances, an analysis of existing operations will indicate deficiencies that may be alleviated or eliminated by the application of new technology. One approach, of course, is to undertake a crash program in-house to develop the improvements needed to restore competitiveness. One problem with this approach, obviously, is the significant time period involved in producing the captive technological improvements. Another is the possibility of making the technological advances required, only to find that someone else's patent coverage precludes application of the advances. Because of these considerations, the acquisition of technology rights from external sources can be a more effective approach.

MORE EFFECTIVE ALLOCATION OF LIMITED RESEARCH RESOURCES

A common activity in the annual planning and budgeting cycle for the process industries and others is the development of a "want list" of technology-related efforts to be undertaken during the coming year. These lists usually include exploratory and basic research efforts, technology development needs, and the ongoing support of manufacturing and marketing efforts. Almost without exception, the needs spelled out in such a list will significantly exceed the resources (personnel, financial, and other) available within the organization. One response to this problem is the cancellation or deferral of a portion of the proposed projects. Another, and sometimes preferable, approach is to obtain some of the targeted technological products by licensing from others. Although there is a financial outlay associated with this option, it may be justifiable on the basis of time savings and risk reduction of proven technology. An acceleration of a few years, for example, in the availability of technology for a new facility can provide financial justification for significant license fees and royalty payments.

TO SUPPLEMENT CAPTIVE TECHNOLOGY

"Home-grown" technology sometimes appears on the scene with some deficiencies that preclude full realization of its commercial potential. For example, the problem may be inadequate catalyst performance or life. In other cases, the deficiency may relate to an inability to produce a product of sufficient purity. Still others may entail excessive energy usage to deliver a product of acceptable quality.

Where the technological product is otherwise superior and a potentially valuable piece of intellectual property with only one "fatal flaw," the acquisition of remedial technology from external sources may be a very profitable salvage operation. Rights to use a particular catalyst, for example, can often be acquired from an outside source on either a purchase or license basis, and thus remedy the competitive deficiency. Acquired product purification technology, such as ion exchange, can sometimes turn a mediocre technology product into a unique asset.

In some cases, it may not be possible to acquire the missing technology component by means of a conventional licensing arrangement. One alternative that has been applied in this situation is a joint licensing venture, wherein licensing income for the improved product is shared with the supplier of the remedial technology. One obvious virtue of this arrangement is the risk-sharing aspect, as well as the minimization of the initial financial outlay by the primary licensor.

FOR SUBLICENSING, RATHER THAN CAPTIVE USE

In most instances, when technology is licensed from outside sources, the objective of the acquisition is captive use to satisfy manufacturing or market needs. In certain cases, however, the technology is acquired specifically for the purpose of sublicensing to others. The sublicensing of the technology may be a stand-alone venture, wherein the sublicensor simply has better access or a more effective distributor network to the ultimate market than the original developer of the technology. This could also be the case where the developer of the technology lacks the critical technical and marketing resources to mount a full-fledged licensing program, but does not wish to relinquish all future rights to the technology.

Sublicensing of another's technology may also occur where it can serve to enhance the applications and sale of a mainline product. A coatings manufacturer, for example, may find it advantageous to acquire a license for new applications techniques or devices, particularly if they fall outside his area of expertise. These devices, mechanical or otherwise, can then be sublicensed to customers for the coatings as part of a total systems marketing effort. Similarly, a foam applications technique for dyes and other additives was acquired by a textile machinery manufacturer for sublicensing as a means of differentiating his product from others in the same field. Here again, sublicensing was the solution to a business need.

FOR FURTHER DEVELOPMENT AND NEW APPLICATIONS (VALUE-ADDED)

There are numerous instances of technology developers and inventors who find themselves with a potentially valuable technological product that requires further technical or market development to realize its full commercial value. In addition, even large organizations will, on occasion, find themselves with technology that does not fit well with their mainline business strategy. In these cases, where there is a shortage of developmental resources or a lack of fit with commercial direction, there can be an opportunity for a mutually profitable technology transfer, wherein value is added by further technical development or marketing effort. Just as many businesses are based on the upgrading of crude or semifinished materials to a finished product for which there is an established market, so too there are opportunities in adding value to semifinished technology products.

Some value-added technology acquisitions occur on an opportunistic basis. An approach by an outside party or an appearance in the literature can serve to identify a potential candidate for acquisition and upgrading to commercial reality. Some entrepreneurial organizations, however, prefer not to rely on some fortuitous circumstance, but rather choose to undertake systematic searches for technology development candidates. These efforts can be carried out by the organization itself or, in some cases, by third-party consultants. We shall take a closer look at search approaches in the following chapter.

Even when a technology product has been fully developed and marketed effectively, there may be an opportunity for adding value by application to a different market. Over the years, for example, technological products developed and produced for the military and aerospace sectors of the economy have found increasing applications in the civilian markets.

TO IMPLEMENT A BUSINESS ACQUISITION

Most licensing or technology transfer agreements are quite specific on the granting of rights to use the technology, and also on the assignability of those rights. Care should be exercised in the course of developing an acquisition of a facility, a market area, or a total business, to define any intellectual property (patents, know-how, trademarks, copyrights) essential to continued operation of the business. Once these critical elements have been defined, existing agreements can be examined and a plan structured to combine intellectual property with the fixed assets in the overall purchase plan.

There is usually no major difficulty in identifying and integrating technology products owned by the seller into the transaction. However, technology products owned by third parties and licensed to the seller for his use may require more effort in order to properly identify continuing obligations (financial, technical, secrecy, etc.) and to ensure continuity. Characteristically, intellectual property transfers will require continued attention and resource application long after the transfer of fixed assets has become a fait accompli. In one acquisition of a Western European petrochemical complex by a British firm, for example, both buyer and seller had technology transfer personnel involved on a continuing basis for at least ten years after the transaction date, dealing with technical support and grantback obligations, as well as related technology transfer matters.

PROBLEMS IN TECHNOLOGY ACQUISITION

Based on this broad range of commercial incentives for technology acquisitions, there would appear to be strong motivation, at least in selected situations, for an active effort in this area. However, technology acquisition is not without its pitfalls and drawbacks. Although not insurmountable, there can, and probably will, be problems in technology acquisition efforts, whether on a "one-shot" basis or as part of a continuing program. Most licensing professionals can cite horror stories about various technology acquisition efforts, wherein the product was not as expected, the price escalated beyond any reasonable expectation, or the transfer of technology was difficult and time-consuming.

As in most commercial transactions, some problems are inherent in the nature of the transaction, and can only be recognized and dealt with, not eliminated. In these situations, understanding and anticipation are key elements in effective management. Still other potential problems, on the other hand, can be significantly alleviated, if not eliminated, by careful planning and technology transfer management. Let us first examine some of the first type of problems, those that are inherent in most technology acquisition efforts.

The first problem of this type that usually comes to mind is undoubtedly the "Not Invented Here" or "NIH" syndrome. The problem, of course, as the name implies, relates to the perception, in those organizations with in-house technology-development capability, that technology acquisition from another source somehow reflects a failure in carrying out their mission. No matter that the acquisition stems from reasons discussed previously that are unrelated to the adequacy of the in-house technology effort; the feeling persists that this new technology is a stranger or, at best, a stepchild in the family.

As previously noted, the problem will not simply "go away" of its own accord. However, some organizations have been able to alleviate the severity of the problem by broadening the mission of the R&D function to include the supply of all technology products, whether developed internally or licensed from outside sources. Skeptics might contend that such an arrangement is equivalent to the "fox guarding the henhouse," but past experience has not supported this premise. Such an organizational arrangement should not be construed as some sort of ploy to sugarcoat an otherwise distasteful move; the professional expertise of the technical function is essential to effective technology search and evaluation.

A second type of inherent problem in licensing technology from external sources is the cost of the transaction. As we have discussed in previous chapters, fees and royalty payments for technology products are quite significant. In addition, they fall in a more highly visible category than the ongoing overhead and payroll costs associated with in-house technology programs. There has always been a running debate as to whether license fees and related expenses are *in addition to*, or *instead of*, in-house R&D expense, and a clear-cut answer is difficult, if not impossible, to come by.

Some have attempted to confront this problem by arguing that the license fees are less than the cost of making the invention in house. Such an argument, based as it is on hypotheses regarding costs for a non-existent technical program, usually generate more emotion than understanding. The one point of agreement is usually the contention that no one really knows in advance the cost for a successful in-house effort.

A somewhat similar, but considerably more conclusive argument is that technology acquired from outside sources can advance considerably the start-up date for the commercial application of the technology. This is usually a more conclusive argument than the relative-cost debate referred to above. There is generally a demonstrable time difference in commercialization between a fully developed technology product available from outside sources and a self-developed product requiring bench-scale, developmental, and perhaps pilot plant work. The financial impact of this time difference is usually quantifiable and significant.

There is a third problem in the acquisition of technology from outside sources. This is the realization that the licensee will usually be at a competitive disadvantage to the licensor by virtue of the required royalty payments, and also the in-depth knowledge gained by the licensor regarding the licensee's manufacturing situation. Some licensees have alleviated this problem by avoiding grantback provisions in the license agreement and mounting an effort to develop and improve the technology further. One U.S. licensee of a foreign monomer technology, for example, employed this technique and appears to have achieved a competitive edge in the use of the original technology, in spite of royalty payments. Such an approach requires a careful comparison of the expected level of licensor capability and support with the cost and effectiveness of the licensee's own efforts. In addition, it may not be possible to negotiate grantback provisions out of the final license agreement.

The other category of problems in technology acquisition could perhaps more properly fall in the "how-to" classification, inasmuch as they involve the techniques of obtaining technology from outside sources. These are the problems of organizing, managing, and implementing a technology acquisition. More specifically, these entail *searching* for the technology, *evaluating* the results of the search, and *managing the transfer* of the technology of choice. Because of the importance of these activities, we shall devote the next three chapters to a consideration of these elements of technology transfer.

12

Technology Search

In most areas of business, the search for new products is an ongoing activity, ranging from mild receptivity to external approaches, on the one hand, to highly organized and focused new product development and acquisition efforts. The same spectrum of new product activity applies also in the area of intellectual property. As we discussed in the previous chapter, there are many factors that can create a need for new technology products. However, it is not sufficient simply to recognize or identify a need. Rather, the identification of a need for a new technological product, in most case, will bring forth a new problem, that of finding a source for the desired technology.

The breadth and depth of a search for a technology product(s) can vary over a wide range, depending on the type of product and the level of definition. In some instances, when a single product and few suppliers are involved, the search can be relatively straightforward and direct. A manufacturer seeking methanol-to-acetic acid technology in all probability, would contact Monsanto, the original developer, as the likely source. (It is recognized that Monsanto has developed business arrangements with others to market the technology, but such an approach will lead to the appropriate source.) On the other hand, an organization wishing to manufacture acetic acid without a predetermination as to raw material may need to approach a host of potential licensors, involving acid processes that are butane-based, ethylene-based, grain-based, and naphtha-based, as well as methanol-based.   Such a search is, of course, an order-of-magnitude more difficult and time-consuming than the single-supplier effort, even with a defined product. Extend this kind of significant activity, then, to a search for technology in a broad product area (such as organic solvents or paper chemicals) and one can visualize yet another step-change in search and evaluation effort.

Considering the complexity of such a search effort in terms of both time and cost, one may be inclined to indulge in some semiarbitrary "short-listing" in order to expedite the effort.   Indeed, to a certain extent, and if practiced by someone knowledgeable in the industry and the technology, some judicious reduction in the candidate list can be accomplished without rigorous technocommercial analyses. However, the reduction in search effort will usually be incremental, and there always exists the possibility of the loss of a potential winner because of premature or overly simplified screening.   Because technology products, as opposed to some tangible conventional products, may have a life, and hence profit impact, expressed in decades rather than weeks or months, the consequences of an inadequate or incomplete search effort can be far-reaching.

Is there, then, little or no hope for expediting this critical search activity?   In most cases, the answer is yes, if one is willing to devote some time and effort to preparation for the search effort.   Such an approach would obviously involve something more than an identification of likely sources, followed by issue of a "standardized" letter of inquiry to each.   Rather, it would involve a careful needs analysis, covering several areas, before making an approach to potential suppliers.   More specifically, preparation and analysis would be desirable in three areas:

o   Technology product characteristics

o   Definition of critical issues

o   Desired commercial features

By themselves, these categories are not particularly descriptive.   Accordingly, some further discussion and a few illustrations may serve to better define this activity.

TECHNOLOGY PRODUCT CHARACTERISTICS

First and foremost, this category entails an understanding, so far as possible, of what one hopes to do with the technology and, hence, what qualities it must have in order to perform the desired function. For example:

o   Does one need a "bare-bones" technology transfer package that one can unilaterally commercialize, or should the desired package include a full line of technical support services, such as engineering, review, training, and start-up assistance?

o   For process technology, what manufacturing capacity is contemplated?

o   Are there any unusual characteristics of the raw material supply that the technology must handle?

o   Are there site conditions that will affect the technology product characteristics required?

o   Are there certain technology characteristics required to make it fit with captive technologies?

o   What special timing requirements apply to the technology transfer?

Some of the above questions may appear to be rather obvious. Nevertheless, most recipients of technology product search inquiries can cite case after case in which inadequate definition of the desired characteristics impeded progress on the search. In some searches, it may be argued that a non-specific, "broad-brush" treatment serves to ensure a wide range of responses. Unfortunately, the multiplicity of non-selective responses can sometimes impair or overwhelm the evaluation effort.

DEFINITION OF CRITICAL ISSUES

Another way of describing this area of preparation for the search effort is the question, "What is important to us?"

Only the one seeking the technology can adequately address such critical issues as:

o   How much risk are we prepared to accept? For example, would we even consider a developed, but not commercially proven, technology product?

o   What are the important economic criteria? By our economic criteria, how important is required capital outlay, as opposed to annual manufacturing costs?

o   Is commercialization time a critical issue, i.e., is market readiness worth a premium?

There have been a number of technology product searches that have been misdirected because of failure to resolve issues such as those cited above. Granted, complete resolution may not be possible before the search effort is undertaken, but even the effort to resolve may provide a clearer focus on the desired product.

DESIRED COMMERCIAL FEATURES

There is a common tendency to defer consideration of most commercial features to some remote date in the future, after technical matters have been addressed. To a certain extent, this is probably appropriate, in order to ensure acceptable product functional characteristics. Some preliminary consideration of desired commercial features, however, can serve to sharpen the focus of the search effort. For example, a desire for geographical exclusivity can serve to limit the search to those suppliers who do not have current licensees in the geographical area in question. Other commercial features that could have a bearing on the search effort include the following:

o   Is there interest in having a licensor take product from our facility as part payment for the technology? If so, the marketing position and capability of the licensor may be a significant search criterion.

o   Are there plans for extensive in-house
     upgrading and improvement of the
     technology? If there are, potential
     suppliers who require grantback pro-
     visions may not be prime candidates for
     the search effort.

o   Does our preliminary knowledge of the
     technological area of interest indicate a
     lot of patent activity? If so, a
     willingness to offer indemnification for
     patent infringement may be a screening
     factor for potential suppliers.

There is obviously a subjective balance that must be
made between an immediate mass mailing of a form letter
seeking a particular technology product and a detailed
analysis of desired product features before undertaking
contacts with potential licensors. The choice is affected by
many factors, including the potential user's understanding of
the product area, his technical and commercial resources, and
the urgency of the effort, among others. In most cases,
there is a clear incentive to do as much preparation as
possible to define the product for which the search is being
undertaken.

SEARCH TYPES AND RESOURCES

Volumes could be, and have been, written on the
subject of product search techniques and methods. The
activity can be subdivided and each element analyzed in
detail, depending on the type of technology product being
sought and the commercial characteristics of the proposed
venture. In our further discussion of technology searches,
however, we shall limit our coverage to two key elements:
the type of search to be undertaken, and some approaches to
identifying potential sources. The following sections describe
some principal types of searches and related activities.

Single Defined Product

Probably the most common type of search, this involves
the effort required to define sources for a technology such
as, for example, a particular chemical or refinery process, a
material with specific properties, or an apparatus to perform

a defined function. As noted earlier in this chapter, there may be only one or two possible sources for a desired technology, and the search effort may thus be considerably simplified. On the other hand, particularly with specialty or performance products, the search for a desired technology can be a long and tedious process.

Although the magnitude of the search effort can thus vary widely, depending on the technology required, all searches start from a common point, that of identifying possible sources. Obviously, the success of the total acquisition effort will depend, in large measure, on how effectively we are able to identify potential sources. Accordingly, we need to consider various approaches to the source identification effort.

For a single defined technology product, the search usually begins with those already using the desired product commercially. At least a portion of these will have their own technology and may be interested in licensing. Others may be licensees and thus have contacts with those who themselves have developed technology. One obvious problem, of course, is the exposure of business strategies to potential competitors that such contacts involve. Technology inquiries are useful elements of commercial intelligence, and their input must be weighed in considering alternative search approaches. Because of this, a third-party consultant is sometimes retained to provide anonymity for the search effort. We shall discuss the use of consultants and other third-party arrangements later in this chapter.

The question, of course, still remains as to methods for identifying those already practicing the technology. For those already associated with a particular industrial sector, as a supplier or customer, for example, the information may be readily available. A new entrant, however, may not have ready access to this information. One source that can provide useful guidance is the *Chemical Week Buyers Guide*, an annual directory of manufacturers and distributors of chemicals published by McGraw-Hill. This directory is particularly useful for smaller-volume, less well-known chemical products, where producers may not be the large multinational firms with well-publicized product lines.

Another useful source is *Hydrocarbon Processing*

magazine's annual petrochemical and refining issues. These publications usually cover only the large-volume, commodity-type processes, and hence include a much smaller number of technologies. However, for those included, there is non-confidential information similar to that outlined in the discussion on "black-box" disclosures in Chapter 5—Product Packaging.

A third, and in many cases, the most effective, area for source identification is through a network of licensing professionals. Contacts made in this manner have the virtue of a clear understanding of the technology transfer aspects, as well as broader coverage in the area of interest. In addition, it usually offers a far more direct approach to the decision-makers in the intellectual property area.

At first glance, tapping into a "network of licensing professionals" may appear to be more of an ideal than a practical, workable concept. However, there is a way to access such a network, and that is through the Licensing Executives Society, whose U.S. headquarters are located at 71 East Avenue, Suite S, Norwalk, CT   06851. A worldwide organization of experienced licensing professionals, the Licensing Executives Society is made up of licensing personnel representing major corporations, law firms with intellectual property practice, engineering contractors, universities, and consultants, among others. For one who has a significant involvement in technology transfer, either in or out, membership and active involvement is highly recommended.

The Licensing Executives Society, as part of its services to its membership, provides a directory of its members throughout the world, sorted according to professional affiliation and geographical location. In addition, it also publishes a compilation of members and their areas of specialization for use by the general membership of the organization. These published resources, together with the personal contacts provided at regional, national and international meetings, can be highly useful in any type of technology product search effort.

Once sources of the desired product have been identified, there remains the initial approach or inquiry for the technology. This can vary all the way from a phone call to a detailed request for proposal, depending on the type of technology, the commercial circumstances, and the technological sophistication of the inquirer. In general, written inquires are more effective, inasmuch as they can be referred and transferred throughout the recipient's organization without distorting the message. When making these inquiries, provide enough information about your proposed venture that the supplier can respond effectively and comprehensively. Be forthright and realistic about your timing requirements. An unrealistic response deadline can turn off a potentially attractive source of supply. Also, be specific and realistic about the scope of the response. Do not request a highly detailed disclosure if all that is needed is "black-box" or evaluation data. Understand that the technology product probably contains proprietary information that the supplier must protect with a secrecy agreement.

As noted in the preceding chapter, know in advance the information required to carry out the technology evaluation effort. In addition, a brief overview of venture schedule and plans can add credibility to the inquiry.

Defined Area Search

Thus far, we have considered the search for specific technology products, usually in conjunction with a defined commercial venture that would use the technology product as a basis. All technologies searches, however, do not fall in this category. A second search category involves a broad-range survey of technologies available in a defined industrial technology area.

There can be a number of reasons for such a broad-range technology search effort. First, it may provide the basis for a significant expansion of business into a new, but related, area, based on new technology acquired as a result of the search. In this instance, the technology could be the prime mover for a new business venture, rather than the means to an end. By proper definition of the area to be searched, one can ensure that the market for the new venture is a familiar one, and that the technology itself, though new, is a reasonable fit with in-house functional resources.

A second driving force for a defined area technology search involves a combination of commercial intelligence and technology audit requirements. A clear picture of new developments in a particular business area can serve to avoid unpleasant surprises from competitors, and also to provide a realistic assessment of the competitiveness of one's own technology. A technology search with this type of primary motivation is sometimes carried out on a continuing, or at least periodic, basis, as opposed to an ad hoc approach.

Third, a defined area search, in certain instances, can be a very useful activity for a supplier to a particular market. By careful monitoring of technological developments in this market, a supplier may be able to detect new developments in the area that could lead to additional or expanded applications for his products. In addition, technological developments could render obsolete certain product applications, and early warning of this trend can provide lead time for remedial action.

Here again, the key challenge is identification of sources, or "knowing where to look." A prime starting point, of course, is the patent literature. Many organizations maintain continuous surveillance of patent activity in a particular sector as an integral part of their technical intelligence effort. There are also commercial services that screen not only the patent activity, but also any kind of published information related to a particular business area. In more recent years, these services have become generally available in an electronic format to users of personal computers.

Trade shows and trade association meetings offer another prime arena for technology searches in a defined area. Even though many of these may be more market than technologically oriented, technically related trends can often be detected in new market developments and in personal contacts made during the meetings.

Finally, the direct approach may be effective in some cases. Direct mailings to key individuals and organizations outlining specific interests in a particular technology area can elicit some constructive responses. As we noted in earlier chapters, there are undoubtedly many technology developers and owners who are actively seeking markets for their products. Be prepared, however, for a broad spectrum of technology product quality.

General Area Search

We come now to the other end of the spectrum of technology searches, so far as selectivity is concerned. This involves a broad search for promising technology products, with little or no constraints on the areas of technology to be searched. At first, this might appear to be a rather impractical exercise, in view of the magnitude of the search area and the diversity of skills required to evaluate and commercialize. As a matter of fact, because of these considerations, this type of search is less common than the two described previously, i.e., *defined product* and *defined area*. However, there are ways of applying some limits to what may seem to be an overwhelming search effort. For example, although not limited to one industry sector, such as metalworking or petrochemicals, there are often defined technology product characteristics, such as state of development, capital requirement, price range, and potential market size, that can serve to focus the effort. Moreover, the field may be reduced further by limiting the potential sources to be monitored.

Even with these limitations, the magnitude of the search and evaluation effort can be intimidating. Why, then, would one undertake a general area technology search? First, the motivation may be closely akin to that of a venture capitalist. There are undoubtedly a number of promising technology products that, with an infusion of capital, marketing, creativity, or a combination thereof, could provide interesting financial rewards. Second, there may be technologies that are simply undervalued or misapplied, wherein acquisition and redirection may result in substantial gains for the acquirer. Third, there may be a rewarding role as a technology broker, matching various parties' needs with un- or underutilized technology products.

Likely source identification for this kind of global search is vastly different from those more structured and focused searches discussed previously. Screening every patent issued worldwide, for example, is probably not a practical technique. Some searchers have found, however, that publicizing a wide-ranging interest in technology product (as in selected technical publications, for example) can bring forth some interesting responses. Great care must be taken, however, to avoid compromising future technology efforts by inadvertent exposure to someone else's proprietary information. Solicitations should indicate clearly that only patent-protected or non-confidential information will be considered and evaluated.

Another type of source that can be very useful in a "broad-brush" search is the inventor innovator meeting. An example of this kind of gathering is the National Innovation Workshop sponsored by the U.S. Department of Commerce at various locations throughout the U.S. Although the main objective of the meeting is to assist technology developers in commercializing and marketing, such a gathering can provide entry to a network of knowledgeable contacts in the field of new technology products.

Also available from the Department of Commerce are various publications listing government-developed technologies. *Government Inventions for Licensing*, for example, is a weekly publication of inventions and licensing opportunities available from various U.S. government agencies. A subscription to this publication may also include the *Catalog of Government Patents*, which summarizes inventions that have been announced throughout the year. In addition, *National Technical Information Service Tech Notes* is available to summarize government technologies on a monthly basis. Information regarding availability and cost of these and other similar publications can be obtained by calling the Commerce Department's Center for the Utilization of Federal Technology (CUFT) at (703)-487-4650. Additional information on these types of sources is included in Section 3 of the Appendix.

A technology source of growing importance and activity is that of university research organizations. In some instances, of course, there are commercial sponsors who have priority claims on the results of the research effort. However, these organizations still produce a number of technology products for which commercial opportunities are actively sought. In addition, there are organizations whose primary mission is to provide marketing and commercial development services for the products of university research efforts. Names and addresses of such organizations, as well as of university research contacts, can be obtained through the Licensing Executives Society directories discussed earlier in this chapter, or from commercial directories.

Finally, and by no means least, there is the third-party approach to the technology search. There are a number of highly qualified individuals and firms who can undertake anything from defined product to general area searches for technology products, and these, too, are listed in the directories cited above. The use of third parties is particularly beneficial if, for commercial or other reasons, anonymity is a requirement. For example, if the search is being carried out, in part at least, for commercial intelligence reasons, a third-party arrangement can be far more effective than dealing directly with competitors, either actual or potential. By the same token, a defined product search for some specific piece of technology can sometimes result in premature exposure of sensitive business plans, unless it is protected by some form of third-party search relationship.

If there is a reason to use a third party such as a consultant or a broker in the search, be prepared to devote sufficient time and effort in preparation to make the relationship work. The third party's effectiveness will be enhanced to the extent that he has a clear understanding of the client's background, concerns, tolerance for risk, and ultimate objectives for the venture. Take care to understand the third party's background and experience in the technology and commercial areas of interest during the selection process. Do not abdicate participation in the search effort. Rather, at the very least, set up periodic reviews and, if necessary, redirection meetings throughout the course of the search effort. There are a number of very competent, highly trained consultants and other third parties who can be very valuable to a search effort, but their contribution can be even greater with proper planning and preparation.

With the approaches discussed thus far in this chapter and by careful use of the sources outlined, a technology search effort, whether for a specific product, a defined area, or a broad-range general area of interest, can usually turn up a number of possible technological products for consideration. As a matter of fact, in some areas of technology, the results from an intensive search effort can tend to be overwhelming in terms of the sheer volume of candidates to be processed. For this reason, we turn now in the next chapter to what may well be the most critical step in technology acquisition—technology evaluation.

13

Technology Evaluation

Ask anyone the principal objective of a technology search, and the reply is usually "to find the *best* technology!" And, in truth, it is difficult to find fault with this commendable goal. It has been said before, but it bears repeating, that the choice of a technology is in reality a marriage, where the consequences are long-term and a mistake can be quite costly. It is safe to say that no one undertakes a search for the runner-up, or second best, technology, and yet this, or worse, is sometimes the result. Moreover, it is interesting to note that, in the case of competing technologies, different searchers, all seeking the "best" technology, will make different selections. Are all but one wrong? The answer, of course, is that they could all be right, depending on the standards and criteria reflecting their individual circumstances and commercial objectives.

The point of all this is, of course, that effective technology evaluation involves something more than the application of a single quantitative measurement to a collection of technology products. There is certainly more than one standard of measurement, and the "right" technology for a particular seeker involves the application of several standards. In addition, the results of these measurements and evaluations must be weighted to correspond to the particular needs of the potential user.

For most technology products, there are at least four standards of measurement, or critical areas, for technology evaluation. These may overlap to some extent, depending on interpretation, but generally fall into the following categories:

o   Technological considerations

o   Economic and commercial considerations

o   Supplier background and experience

o   Supplier technology transfer capability

No attempt has been made to list the categories above in order of importance, inasmuch as this will vary with the type of technology and the needs of the user. A technically sophisticated user, for example, may not attach as much significance to supplier background as would a user with less internal technological resources. However, it is highly likely that nearly any potential user will apply all four criteria in one way or another.

TECHNOLOGICAL CONSIDERATIONS

The heart and soul of this category is the familiar question, "Will this technology product work 'as advertised'?" We have discussed at some length the kind of technological data usually included in preliminary "black-box" disclosures. Inasmuch as these data are more often than not the basis for part of the evaluation, a close examination and understanding is essential. For example, there should be a clear understanding of the source of the data in the disclosure. Is it based on sustained operation of a commercial facility, or is it an average of a series of experimental runs? Was the raw material requirement derived from measured consumption or from analyses of various process streams? If the latter, what methods of analysis were applied? Are there special quality requirements for feedstocks in order to achieve the stated consumption? Obviously, commercially demonstrated evaluation data are desirable, but not necessarily essential. However, scrutinize carefully those data described as "better than demonstrated commercial performance because of laboratory-based process improvements." The claim is probably quite sincere and well-intended. Nevertheless, the shortfall in performance going from laboratory to commercial performance is a fact of life that can turn great expectations into severe technology transfer problems.

Take time to explore carefully with the potential supplier the kind of ongoing improvement and support programs under way in this particular technology area. Although probably unwilling to disclose proprietary technological data, most licensors will describe the kinds of planned activity devoted to improving the technology product and will provide an overview of the past efforts in this area. An understanding of the improvements made by the licensor in his technology over, say, the past five years may provide some insight into the general level of technical support to be expected.

Establish whether the technological product in question is completely owned and controlled by the licensor. Is a part of the technology, such as a catalyst or a separations technique, under control of a third party and thus subject to a separate series of evaluations and negotiations? Establish as early in the evaluation as possible how many technological products there actually are in the package being sought.

Also, even though it might appear premature, raise the question of guarantees on technology performance. As we have discussed previously, the significance and value of guarantees can very widely, depending on the needs and technological perspective of the licensee. Nevertheless, the licensor's willingness to provide meaningful guarantees can provide some interesting insights into his confidence in the technology. Would the licensor guarantee the performance set forth in the preliminary disclosure, or does he advertise the disclosure data as "expected" results, with guarantee values set somewhat lower?

ECONOMIC AND COMMERCIAL CONSIDERATIONS

Although in some parts of the world a technology-driven venture may be undertaken for reasons of national policy or to fulfill the goals of a five-year plan, the economic and commercial characteristics of a technology product nearly always weigh heavily in the evaluation effort. Almost irrespective of the economic system under which the venture will operate, the success of the effort will be expressed in some form of financial or economic terms. The capital cost of the facility, for example, is a significant consideration regardless of the economic system in place. The same is true of manufacturing costs and related components, such as raw material and energy consumption.

Because of this, economic and commercial considerations are an important part of any evaluation effort. Some form of base-case economic analysis is usually carried out, whether by return on investment, discounted cash flow analysis, simple payback, or some other technique. As a matter of fact, the content of most preliminary or "black-box" disclosures is tailored to provide sufficient data for such analyses. For example, subject to the disclaimers cited above, one can quantify the annual manufacturing cost by using the raw material, energy, and direct labor requirements contained in the disclosure, applying one's own unit costs, and inputting appropriate overhead factors.

The principal problem in carrying out economic analyses with preliminary non-confidential disclosure data lies in the development of capital cost. Most disclosures of this type include an estimate of the capital cost for a facility using the technology, based on some nominal production capacity. In addition, some disclosures will provide a very approximate method of adjusting the estimate of capital cost to other capacities over some limited range. The difficulty lies in ensuring, first, that the estimate properly represents the cost of the potential licensee's facility under his site conditions, and second, that the estimate is consistent with similar estimates obtained from other licensors for competing technologies. For example, there is usually no way of establishing labor costs and productivities, purchase costs of major equipment items, or allowances for engineering costs, field overhead, or contingencies.

This problem could be solved if one had access to detailed equipment lists and descriptions, flow diagrams, and other process data from each potential supplier, and proceeded to develop capital cost estimates for each competing technology from this common basis. Unfortunately, licensors will seldom supply this level of detail, inasmuch as it entails essentially full disclosure of proprietary details of their technology at a very early stage.

Several approaches have been used to confront this problem. In some cases, particularly where there is only one supplier, a paid disclosure is obtained under secrecy agreement, and the information contained therein is used as the basis for an estimate of capital cost. Alternatively, a third party, such as a contractor or consultant, can be retained to receive disclosures under confidentiality agreements and to perform the necessary estimates of capital costs on a consistent basis. With this arrangement the third party will be prohibited from disclosing details of the technology to his client, and one must thus be willing to rely on the capability of the third party to ask the right questions and to perform the necessary analyses. Finally, in the absence of a third-party arrangement, one can go back to the licensor(s) and request additional information that can permit a better evaluation and comparison of capital costs related to the technology. For example, most licensors will provide a breakdown of direct material, labor, engineering, overhead, and contingencies included in their estimate. In addition, it may be possible to obtain information regarding labor costs and productivities used by the licensor in preparing his estimate. Although this information, of and by itself, does not permit the preparation of budget-quality estimates, it does allow the evaluator to reflect his own site, engineering, and construction parameters and to develop a more consistent comparison among competing technologies.

The base-case economic analysis of the technology product under consideration is, of course, a primary objective of the evaluation. It is important to remember, however, that the conditions assumed for the base-case analysis, such as product sales prices and volumes, feedstock prices, and capital costs, are seldom predictable with any great accuracy. Accordingly, it is usually desirable to include in the analysis the sensitivity of the results to variations in these conditions. The return on investment for one technology, for example, may be considerably more resistant to sales price deterioration than a competing technology. Sensitivity analyses of this type, combined with a prospective user's judgement regarding future uncertainties, can provide a useful form of risk analysis for the technology selection process.

Some evaluators will extend the sensitivity analysis by developing several scenarios of future trends and evaluating the economics of competing technologies under each of these scenarios. These would include hypotheses regarding future trends in feedstock and energy availability, new product applications, and other factors affecting the venture. Clearly, this type of evaluation requires a judgmental balance between our ability to project future trends and the amount of detailed case-by-case analysis warranted for the evaluation effort. There have been classic cases of overkill wherein a multiplicity of scenarios and analyses have served to obscure, rather than clarify, the evaluation effort.

SUPPLIER BACKGROUND AND EXPERIENCE

It is almost axiomatic that the ideal supplier is one who has extensive background and experience, not only in the commercial practice of the specific technology itself, but also in the broader area of technology development, practice, and support. As we discussed in the technology section, the plans for ongoing support of the particular technology are quite important. Of even more fundamental importance, however, are the general capability and perspective of the potential supplier regarding the role of technology in his overall business strategy. In the supplier's organization, is technology regarded as a "necessary evil," or is the organization technology-driven, based on the extent of new product development, patent activity, and general level of support? Although there is no one "right" attitude toward technology, a technology-oriented organization will usually be more inclined to follow through with improvement efforts.

In a number of instances, the supplier of technology, so far as direct contact is concerned, is an engineering contractor, who is acting on behalf of the owner, developer, and practitioner of the technology. Although these arrangements usually produce satisfactory results, never lose sight of the fact that the technology perspective of the owner/developer is probably more important than that of the contractor or other representative. Even the most dedicated contractor cannot overcome a lack of commitment and support on the part of the owner, and the results of such a difference in motivation can be extraordinarily harmful to the commercial health and well-being of the licensee. Insist on direct access to the owner/developer, and draw your own conclusions regarding his interest, capability, and dedication to technological activities in general. Although the attitude toward technology is a somewhat subjective element of corporate culture, past achievements and the organizational structure can indicate a great deal about the perspective.

TECHNOLOGY TRANSFER CAPABILITY

The technology may be quite competitive, the licensor may have a strong technical orientation, the economic and commercial aspects may be very favorable, and yet the licensing effort and the venture it supports may encounter serious problems. One reason for this phenomenon is shortcomings in the fourth evaluation category, that of technology transfer capability. Technology transfer capability, as used in this discussion, refers both to the know-how required for effective technology transfer capability and to the motivation and willingness to support the licensing effort. Without this kind of capability, even the strongest technology with world-class economics can create serious problems for both licensor and licensee.

How does one evaluate this capability? In spite of the fact that this is a somewhat subjective quality, there are some clues that can be detected at an early stage of the discussions. At the first meeting, are there full-time licensing professionals involved, or are the owner's representatives all "conscripted" from other functional assignments to deal with the inquiry on a ad hoc basis? There should be a "mix" of licensing and technical personnel to ensure an effective technology transfer effort.

The amount of organization and preparation for the meeting can also be a reflection of technology transfer capability. Is there an agenda clearly laid out in advance and do all the licensor representatives clearly understand their role in the proceedings? Does there appear to be a workable plan for the technology transfer sequence, should the negotiations continue? Is there effective use of handouts and other visual aids in order to make the best use of limited discussion time?

Adaptability is another characteristic of technology transfer capability. Is there a "take-it-or-leave-it" attitude on the part of the licensor representatives, or is there some flexibility indicated with regard to the prospective user's technology transfer needs? Most knowledgeable licensors have discovered through experience that one rigid technology transfer modus operandi will not fit the needs of every licensee.

Finally, and probably most importantly, find out about the licensor's track record in technology transfer. What technologies has he licensed in the past, and to whom? Is he willing to supply names of responsible licensee managers and professionals with whom he has dealt in past technology transfer efforts? A key factor in the decision to license monomer technology from a German licensee, who did not himself have an operating unit, was the contacts with his previous licensees, and their 'testimonials" as to the licensor's technology transfer capability. A good track record in past licensing efforts is no guarantee of future success, but it can provide a strong indication of expertise and professional competence in technology transfer transactions. Be particularly alert for those organizations whose past performance shows that their principal technology transfer concern is the diversion of resources from other activities.

14

Managing the Transfer

Once an effective search has been made and the results of the search thoroughly evaluated, there still remains the task of managing the transfer of technology from the chosen source. Although the proper choice of licensor can serve to alleviate the many problems inherent in technology transfer, it cannot substitute for effective management of the transfer effort itself. Regardless of the experience of the licensor and irrespective of the involvement of third parties, such as an engineering contractor, there is no substitute for active management of the transfer effort by the licensee.

One of the principal problems in the effective management of a technology transfer is that of addressing the major issues at a sufficiently early stage of the transfer effort. All too often, very substantive issues become "cast in bronze" early in the transfer process as a result of the licensing agreement itself. Even after the agreement is executed, there is a need for earlier resolution of significant issues than if one's own technology were being used. For example, it is often necessary to address key staffing decisions earlier in the project in order to ensure that the right individuals are sufficiently involved in preliminary disclosures.

The key elements in managing the transfer fall generally

160

into two categories. The first involves those issues that are included in, or related to, the licensing agreement. Many of the important management areas in the overall technology transfer effort are completely fixed by the terms of the agreement. Because of this, the management effort must begin no later than the development of the agreement.

The second category involves those activities subsequent to the agreement that are required in the implementation of the transfer. Even the most carefully worked out licensing agreement cannot guarantee an effective transfer effort unless there is a clearly developed plan covering all aspects of the effort from the execution of the agreement to a successful start-up and initial operation.

This whole area of transfer management would be considerably more straightforward if there were a set of rules that set forth the "right" provisions to include in a licensing agreement and the "correct" procedures in implementing the transfer following the agreement. However, these provisions and procedures will vary considerably, depending on the technology, the parties, and the business strategies involved.   We can identify, however, some guidelines, or a checklist, that can be applied as management tools in the two categories of transfer management identified above.

TECHNOLOGY TRANSFER MANAGEMENT IN THE LICENSE AGREEMENT

Ensure that the technology disclosure arrangements in the license agreement are compatible with your project execution and schedule requirements. A common problem in many license agreements is the understandably different objectives of licensor and licensee regarding timing for disclosure of needed technical data. The licensor wants to ensure that his technology is not prematurely exposed, and thus would like to avoid detailed disclosures until after a license agreement has been executed by both parties. The licensee, on the other hand, has a serious need for a certain level of understanding of the technology in order to prepare a project proposal for review and approval by his management. Such a project proposal will usually include elements such as the capital cost of the facility, which must, in turn, be based on a reasonably clear understanding of the technology. This clear understanding of the technology can be gained, of course, by execution of the license agreement; however, execution of the license agreement could entail sizable financial commitments before management approval of the project—a tactic that is usually not looked on with much enthusiasm.

An alternative approach is to develop the project proposal using only the information available on a non-confidential preliminary disclosure basis. This, too, has its drawbacks. A vivid recollection is that of a U.S. Gulf Coast licensee who chose to use the preliminary disclosure estimate of 9 million dollars for the battery-limits capital cost of the plant using the licensed technology. The final cost of the facility, with no change in capacity or other variables, was 27 million dollars. Although there were factors other than lack of technical information contributing to the overrun, a clearer picture of the technology and its equipment requirements would have significantly reduced the discrepancy.

There are some approaches that can be taken to address this dilemma. First, to the extent possible, and starting back from the desired facility completion date, outline a schedule of information and data needed to satisfy project evaluation and approval requirements. Do this as part of the preparation for negotiating a license agreement. Although the licensor will continue to be concerned about premature disclosure of his technology, his receptivity to proposals for different disclosure schedules will be far greater before or during negotiations than after an agreement has been executed.

There are various approaches that can serve to provide the needed information and yet protect proprietary technology. A paid disclosure under a secrecy agreement can serve in many instances to provide a significant portion of the needed data without divulging all key elements. If this is not possible or practical, consider a third-party evaluation. In this approach, a third party, such as a consultant or a contractor, receives a disclosure under a secrecy agreement that prohibits transmittal of technical data to his client. The third party then prepares the necessary evaluation, including such items as estimates of capital costs and economic analyses, technical judgments, etc., for submittal to his client. Although these procedures may seem a bit cumbersome, they do serve to provide the necessary information in a timely fashion while still protecting the intellectual property.

Consider implications of planned technical support programs on the desirability of grantback clauses in the agreement. The fact that a technology has been purchased does not preclude the initiation of an extensive technical support and improvement program by a licensee. Technologically sophisticated licensees, in particular, have found it advantageous to undertake process improvement programs on licensed technologies. These programs, however, may make it significantly less desirable to have a grantback provision in the license agreement, wherein rights to technical advances by a licensee are granted back to the licensor and his other licensees. The issue here is not the desirability (or lack thereof) of grantback clauses per se, but rather the desirability of establishing a position on the issue before having to execute the license agreement. Attempting to define the desirability of technical support programs three or four years in advance may seem premature, but the additional negotiability of the grantback issue can make the effort worthwhile.

Wherever possible, reference payments or other licensee obligations under the agreement to progress in construction or implementation of your venture, rather than to the mere passage of time from agreement execution. For large ventures, in particular, schedules are subject to slippage and delays because of management, marketing, or financial factors. Checkpoint events, such as groundbreaking, mechanical completion, or capacity operation, are definable and have the virtue of matching payments to the licensor with the licensee's progress toward income production from the venture. Although it may not be possible to agree on relating all payments to licensee-controlled events, even a partial move in this direction can serve to alleviate the financial impact of unavoidable delays.

Ensure that the impact of secrecy obligations in the licensing agreement are understood and accepted throughout the licensee organization. There is, unfortunately, a tendency to look upon the language in secrecy agreements as "boiler-plate" affecting only the transaction covered by the agreement. In some cases, however, secrecy clauses could compromise work under way in related technology areas of the licensee's organization. Obviously, the secrecy obligations, as well as the rest of the agreement, must be reviewed by an attorney for the legal implications. In addition, it may be useful to ensure that there is a general awareness throughout one's technical organization of plans to incur the secrecy obligation. In this way, the pre-existing knowledge in similar areas can be documented, if necessary, and future problems related to the origin and ownership of certain bodies of knowledge alleviated or eliminated.

If the transaction represents the first license granted, explore opportunities for significant concessions from the licensor. In certain cases, the willingness to license may be the major, and the only, concession possible. However, for most active licensors, the existence of a successful licensee installation can be a powerful marketing tool, justifying significant incentives. Price or royalty rates are certainly an obvious area for concessions, but there are also other possibilities. As the first licensee installation, and thus an integral part of the ongoing marketing effort, a case can be made for some sharing of revenues from future licensees. This would be particularly appropriate if the licensor has no operating facility suitable for "show-and-tell" efforts in marketing the technology. If this kind of revenue-sharing arrangement were not workable, one might well explore the possibility of some form of geographical exclusivity, as a protection from future technology marketing efforts in the area. Certainly the additional risk inherent in being the first licensee of a particular technology, along with one's possible future role in the licensor's marketing efforts, warrants tangible incentives.

If not the first licensee, seek early access and communications with other licensees of the same technology. Although there may be competitive or proprietary obstacles to unrestricted access, an understanding of others' problems in managing the technology transfer from the same licensor can be extraordinarily helpful in avoiding repeated mistakes. Be prepared, however, to reciprocate by offering access to your facility when completed and operating.

Some licensors offer periodic licensee meetings wherein technical and operating improvements are communicated and discussed to the mutual benefit of all. If this is the case, be sure there is a clear, documented understanding of the point in the technology transfer transaction at which you will have access to this significant technical resource. There is sometimes an attempt to limit a licensee's access until he is actually practicing the technology. Even if this is the case, try at least to obtain copies of papers and reports presented at past meetings. The value of this information, in some cases, can be significantly greater in the planning and design stage than in the operational phase, and thus even partial early access can be of considerable value.

TECHNOLOGY TRANSFER MANAGEMENT AFTER THE AGREEMENT

Ensure early and continued involvement of technical functional specialists in the technology transfer effort. All too often, there is a strong belief throughout the licensee's organization that the licensing of technology essentially eliminates the need for in-house technical resources. Add to this the time and expense involved in visits to licensor and other licensee facilities, and it is probably not surprising that there is a tendency to skimp on resources. However, if there is one key activity in the overall technology transfer effort, it is the effective gathering and application of information from the licensor and possibly other licensees. It is seldom, if ever, that one or two individuals can provide the necessary breadth of experience to ask the right questions about every specialized element of the technology. It should be obvious, but the execution of a license agreement does not make technology transfer happen; the activity must be managed, and an essential element of the management effort is the prompt and selective assignment of the appropriate technical resources. Standard or generic written disclosures from the licensor are useful, but they cannot anticipate all the licensee's specialized needs and requirements. Early involvement of these critical skills and an active role by the licensee in the transfer process, can provide low-cost insurance against severe future operational difficulties.

Avoid third-party barriers to technology flow.   In some cases, the actual design of the facility employing the technology will be by an engineering contractor.   In addition, the contractor may have an ongoing relationship with the technology licensor wherein he (the contractor) acts as the primary representative of the licensor.   Such an arrangement, of course, can serve to enhance the contractor's marketing position considerably, and also minimize resource requirements for the licensor.   There is nothing wrong with such arrangements, and there have been a number of successful ventures based on them.   Nevertheless, there is no substitute for direct contact and communication between licensor and licensee.   The licensor will usually have the perspective of a technology developer and practitioner and, as such, will be able to furnish insights and answers that may not be available elsewhere.   In addition, a clear indication of interest in the licensor's involvement can lead to a more effective working arrangement, with beneficial results for all the parties.

Establish defined channels of communication between licensee and licensor.   This may appear obvious, but is often overlooked in the urgency of getting the technology transfer completed.   There is sometimes a tendency, particularly after the first few meetings, for individuals with special interests to communicate directly with those of similar backgrounds in the other organization.   Nevertheless, and notwithstanding the discussion in the previous section regarding organizations and communications barriers, it is usually beneficial to establish one individual as the focal point for all communications between the parties.   This does not preclude, for example, phone conversations between specialists, but a brief note documenting the communication is advisable. Written communications will flow through this individual, and a record will be kept of pending replies so as to facilitate follow-up on sensitive items.   As in most procedural issues, implementation must be tempered with common sense in order to ensure that the primary objective of effective technology transfer is being met.   However, past experience has indicated that an orderly communications process can enhance, rather than inhibit, the technology transfer process.

Part IV

Appendix

In any book that addresses a field as extensive and varied as technology transfer, there is nearly always a dilemma as to the inclusion of useful resource and reference material. Should such material be included in the main body of the book where it will be most accessible, but may disrupt the logical organization of the primary material? Or rather should it be collected with other similar material in an appendix or addendum, with appropriate references from the main text?

For the most part, the choice has been the latter option, that of a separate appendix section. This permits the inclusion of more comprehensive data and information, where appropriate, as well as a broader range of treatment to reflect the global nature of technology transfer.

The appendix is organized in the same sequence as the text of the book itself. The first section deals with the general area of intellectual property, with particular emphasis on patent protection and its application throughout the world. Although not intended as a substitute for the professional involvement of a patent attorney, it is believed the material will be useful in understanding the characteristics of patent protection in various countries of the world, and thus helpful in protecting the value of technological assets.

The second section of the appendix relates to international marketing, or out-licensing, of technology products. The principal emphasis in this section is on giving the reader an understanding of the law, regulations, and procedures relating to export marketing, particularly those applying to technological products. Included is an overview, from the U.S. Department of Commerce, of the procedural aspects of complying with export regulations, as well as selected excerpts from the regulations. This area is one that is often overlooked in marketing technology, leading to embarrassment, costly delays, and in some cases litigation.

The third section of the appendix provides a brief sample of one search technique for technology products. As noted in that section of the main text, the identification of sources for specific products is usually somewhat more direct and straightforward than the more generalized non-specific, multiproduct area that is addressed in this section of the appendix.

Section 1

Patent Laws for Various Countries

As discussed previously in Chapter 2, a strong contributor to the value of intellectual property is the capability of being protected by patent coverage. The "monopoly rights" conferred by patent protection enhance the value of a technology product to a potential licensee and provide a substantive justification for significant license fees and royalty payments.

Patent protection, however, will obviously vary from country to country. In view of the global nature of technology transfer, it behooves anyone active in the field to have a rudimentary understanding of the treaties and agreements governing international patent protection, as well as a general awareness of differences among individual countries in patent practice and procedures.

One of the clearest treatments of this complex subject appears in a paper by Joseph M. Lightman, appearing in an April 1985 U.S. Department of Commerce publication, *Foreign Business Practices*. The following material is taken from that paper, which is entitled "*Foreign Patent Protection: Treaties and National Laws*."

"The Patent Corporation Treaty (PCT) is currently adhered to by the United States and 34 other countries. The PCT is administered by the World Intellectual Property Organization (WIPO), which has published a *PCT Applicants Guide* containing detailed information on the Treaty for those interested in filing international applications, a *PCT Gazette*, and brochures containing a text of the PCT, its regulations, and Administrative Instructions. These publications may be purchased from WIPO at 34 Chemin des Colombettes, 1211 Geneva 20, Switzerland.

Basically, the treaty provides centralized filing procedures and a standardized international application format. Under the PCT, a U.S. national or resident may file an international patent application of the U.S. Patent and Trademark Office (PTO) and designate in that application the member countries in which he desires patent protection. This filing has the same effect as if that person had filed several or many individual applications for the same invention in those member countries. After filing, the application is subjected to a search of the prior art by an international searching authority, which, for the U.S. applicant, is also the U.S. Patent and Trademark Office. The applicant, when he receives the International Search Report, can then decide whether he wishes to continue with the national patent granting procedure of his application in the countries he has designated. The PCT benefits the U.S. applicant by enabling him to make a single filing in the United States of an international application in English and according to a uniform format. This should minimize the expenditure of time and money, particularly with respect to such formal requirements as certifications, consul stamps, and other foreign legal procedures.

The U.S. applicant is also provided additional time (up to 20 months, instead of the ordinary 12 months) within which to submit translations and national fees to foreign countries. During this time he will have available the International Search Report containing prior art citations (patents and other published technology which might disclose the invention) which may further aid him in deciding whether to proceed in one or more foreign countries.

As a long range benefit, the treaty should provide a focal point for continued cooperation among the world's patent offices toward the improvement of patent practices, to the advantage of U.S. exporters. The treaty is also designed to benefit developing nations by providing for the establishment of information services to facilitate acquisition of technology and technical information. Additionally, it calls for a committee to organize and supervise technical assistance programs to aid developing countries in improving their patent system, so also should their market potential become increasingly attractive for U.S. investment of foreign technology, know-how, and capital.

The PCT makes no substantive changes as to patentability of an invention in an individual member country. The patent grant ultimately rests with the member country. The main propose of the PCT, then, is the facilitation in the filing of patent applications in other than one's own country. It deals with procedural matters and does not result in the issuance of international patents.

Under the European Patent Convention (EPC), any party, including a U.S. national, may file an application with the European Patent Office (EPO) in Munich, Germany, or its branch office at The Hague, Netherlands, designating those member countries where patent protection is desired. First, the application is examined for formalities by the EPO. If it is found acceptable, a search is then conducted.The application is published 18 months after the official filing date. The search report, when it completed, will accompany the application. Within 6 months of publication of the search report, the applicant must file a "request for examination"; otherwise the application will be considered withdrawn. If such a request is made, the application is then examined by the EPO for novelty and inventive merit, resulting in either refusal or grant of a patent. The grant of a patent is published in the *European Patent Bulletin*, and for 9 months thereafter anyone may file an opposition. If an opposition is filed, the application will be re-examined and the patent either sustained, modified, or revoked. Appeal procedures are available regarding opposition decisions. Duration of the European patent in the countries designated by the applicant is 20 years from the effective filing date.

An EPC patent has the same effect in the designated countries as if the patent had been granted by the individual member country in terms of uniform examination and on questions of validity. Questions of infringement, however, are left to member states under their national laws. Thus the EPC introduces a step toward progressive unity in the filing of applications and grants of patents, but does not go so far as to remove all rights from member countries relating to patents. Belgium, France, Federal Republic of Germany, Italy, Luxembourg, Netherlands, Sweden, Switzerland, and the United Kingdom are now members of *EPCA Guide for Applicants—How to Get a European Patent* may be secured by writing to the European Patent Office (EPO), Motorama Haus, Rosenheimer Strasse 30, Munich, Germany.

The centerpiece of international treaties on patent rights is still the oldest and most important of those in existence—the Paris Convention of the Protection of Industrial Property (commonly known as the "Paris Convention") founded in 1883. The United States has been a party since 1887. Presently, 88 countries are members. The Convention applies not only to patents, but also to trademarks, industrial designs, utility models, trade names, and, under the 1967 Stockholm Revision, to inventors' certificates. The main provisions concern national treatment and the right of priority.

Under the national treatment provision, the Convention provides that, with regard to protection of the aforementioned types of industrial property, each member country must grant the same protection to nationals of other member countries as it grants to its own nationals. This provision guarantees that foreign applicants will be treated at least as well as domestic applicants in pursuing protection of their industrial property rights. Under the right of priority provision, on the basis of a regular application first filed in one of the member countries, the applicant may, within a certain period (12 months for patents), apply for protection in any of the member countries and have such later filed applications regarded as if they were filed on the same date as the first application.

The convention also contains provisions designed to protect patent owners against arbitrary forfeiture of their patents if not used or worked. It also establishes the principle of independence of patents meaning that, once a patent has been granted, its subsequent revocation or expiration in the country of original filing does not affect its validity in other countries. The Convention also provides safeguards against invalidation of a patent merely because the patented product was imported into the country of destination.

The United States also belongs to an Inter-American Convention on Inventions, Patents, Designs and Industrial Models signed at Buenos Aires in 1910. The Convention, adhered to by 12 Latin American countries, adopts the principles of national treatment, right of priority, and independence of patents along the lines of the Paris Union Convention. Most countries, including those that have become independent since the end of World War II, have patent laws. There are only a few without a system of patent protection. Many former colonies, now independent countries, provide patent protection only on the basis of a patent first acquired in the former parent country. Examples are Ghana, where a "confirmation patent" is issued, based on a patent first acquired in the United Kingdom, and Burma, where Indian patents are recognized as being in force.

There are certain countries that provide, in addition to regular patents of invention, so-called "introduction," "revalidation." or "importation" patents. These can be applied for on an invention already patented elsewhere by the same patent owner or, after a period of time, by a local national. Such patents expire at the duration of their basic foreign patents. Their purpose is to permit an invention to be introduced and protected, notwithstanding its prior patenting in other countries.

The U.S.S.R., Bulgaria, Poland, Romania, Albania, and Algeria provide for issuance of so-called "inventors' certificates," as well as patents. The inventors' certificate system is used extensively in the U.S.S.R. Under its procedures, an inventor offers his invention outright to the state, which assumes its ownership and exclusive use. If used by the state, it entitles the inventor to a cash reward or other specific benefit. In such dual system countries, the inventor generally has the right to choose between applying for an inventors' certificate or a patent. Local inventors in Eastern Europe generally apply for inventors' certificates; foreigners apply for patent rights because of certain impracticalities in acquiring inventors' certificates."

On the following pages is a country-by country summary of world patent laws. It is emphasized that this summary is provided only as an overview of the differences in local protection practices. Consultation with a patent attorney is highly recommended in order to ensure proper interpretation and updated versions of the applicable law.

WORLD PATENT LAWS: COUNTRY BY COUNTRY SUMMARY

The material in this section was taken from an article by J.M. Lightman entitled "Foreign Patent Protection: Treaties and National Laws," appearing in a DOC publication *Foreign Business Practices*.

Afghanistan

No patent law. Some common law protection available for inventions and designs against imitation.

African Intellectual Property Organization

Member countries: Benin, Cameroon, Central African Empire, Chad, Congo, Gabon, Ivory Coast, Mauritania, Niger, Senegal, Togo, and Upper Volta.   Inquiries and applications should be directed to the Office African de la Propriete Intellectuelle (DAPI) located in Yaounde, Cameroon.

Invention patents valid in all member countries 20 years after application. Prior publicity anywhere prejudicial. No novelty examination. No opposition provision. Compulsory licensing possible 3 years after patent grant or if working interrupted for any 3 year period. French patents in force prior to dates of independence of the various countries may receive protection for 20 years from application filing date if revalidated with DAPI before March 31, 1967.

Albania

Invention patents valid 15 years from application date; inventors' certificates also granted. Chemical manufacturing processes patentable, but not chemicals; medical and some biological inventions eligible only for certificates. Prior publication or use anywhere prejudicial. Novelty examination. Opposition period 3 months. No provision for working. Compulsory licensing possible.

Algeria

Invention patents valid 20 years from application filing date. Confirmation patents valid 10 years from filing date of foreign patent upon which based. Inventors' certificates also granted. Prior publicity anywhere prejudicial. No novelty examination or opposition. Confirmation patent must be worked 1 year from grant and not discontinued for more than a year; otherwise can be cancelled. Compulsory license of invention patents possible 3 years from grant or 4 years from application date if not adequately worked. French patents valid in Algeria on July 3, 1962 remain in force if continuously worked by Algerian enterprise.

Antigua

Patents valid 14 years from application filing date. Confirmation patents, coterminous with U.K. patents also granted; must be filed for within 3 years of latter. No novelty examination. For independent patents, public use in Antigua prejudicial. No working. Compulsory licensing possible.

Argentina

Invention patents granted for 5, 10, or 15 years; 15 years after grant only for important inventions. Patents of importation good for up to 10 years. Pharmaceutical manufacturing processes patentable. Prior publication anywhere, grant of foreign patent (except Argentine import patent), or public use in Argentina prejudicial. Novelty examination. No opposition provision. Working required 2 years after grant, not to be interrupted for any 2-year period. No compulsory licensing provision. Importation or advertised offer of sale may constitute working.

Australia

Invention patents valid 16 years after application, renewable up to 10 years where inadequately remunerated. Prior publication, public use, or disclosure in Australia prejudicial. Novelty examination. If no examination requested within 5 years of application filing date, application will lapse. Opposition period 3 months. Compulsory licensing possible 3 years after grant if inadequately worked; revocation possible 2 years after first compulsory license. Patent registration should be marked on product.

Austria

Invention patents valid 18 years after application. Prior published description anywhere, or use or exhibition in Austria prejudicial. Novelty examination. Opposition period 4 months. Compulsory licensing possible 3 years after grant or 4 years after application, if inadequately worked.

Bahamas

Patents granted before June 1, 1967 (new Act effective date), valid 7 years, renewable twice for 7 years each time. Under new Act, invention patents valid 16 years from application filing date. Publication, public use, or knowledge prejudicial. No novelty examination, opposition, compulsory licensing, or working requirement.

Bahrain

Patents valid 15 years from application filing; renewable 5 years. Usually granted as revalidation of foreign patents for term-duration of latter. No examination, working, or compulsory licensing provisions.

Bangladesh

Patents granted in Pakistan before independence date (March 25, 1971) and still in force then can be maintained for their original duration in Bangladesh upon payment of necessary fees. New applications may be filed in Dacca. Patents valid 16 years from application date. Novelty examination. Four month opposition period. Compulsory licensing possible.

Barbados

Invention patents valid for 14 years from application date, renewable for 7 years. Provisional protection available 9 months. Public use in Barbados prejudicial. No novelty examination. Opposition period 2 months. Compulsory licensing possible.

Belgium

Invention patents valid 20 years after application; patents of importation valid up to 20 years. Patentable inventions must be industrially or commercially workable. Prior commercial use in Belgium, or patenting or publication anywhere prejudicial except for import patents. No novelty examination; form only. No opposition provision. Working required 4 years after application filing (1 year for countries not party to Paris Union) not to be interrupted for any 12 month period. Compulsory licensing provisions.

Belize

Patents valid 14 years from application filing date, renewable for 7 to 14 years. Coterminous with prior corresponding foreign patent, if one exists. Confirmation patents based on and coterminous with U.K. patents granted, if applied for within 3 years. No novelty examination, working or compulsory licensing provisions.

Bermuda

Invention patent valid 16 years from patent grant, renewable 7 year periods. Patents also available as confirmation of U.K. patents, if applied for within 3 years of latter's grant date. Coterminous duration with latter. No novelty examination. No opposition for independent patents; 2 months for confirmation patents. No working provisions but compulsory licensing possible.

Bolivia

Invention patents valid up to 15 years after grant, including renewals; confirmation patents valid up to 15 years. Prior knowledge, description, working or non-patent publication anywhere prejudicial. Foreign patent not prejudicial, if application filed within 1 year of foreign application. No novelty examination. Published twice in 1 month at 15-day intervals for opposition. Compulsory licensing possible 2 years after grant, if inadequately worked, or if working interrupted for any 1 year period. Importation or advertised offer to license may constitute working.

Botswana

Confirmation patents based on prior registration in South Africa. U.K. patents automatically in force, recordation unnecessary.

Brazil

Present law effective December 31, 1971. Patents granted before then valid for terms stated in patent grant. Invention patents are valid for 15 years from application filing date. Use or publication anywhere prejudicial. Application published 18 months from earliest priority or filing date. Applicant can request examination within 24 months; otherwise application deemed abandoned. Opposition period 90 days. Working required within 3 years after grant and not interrupted for longer than 1 year, otherwise subject to compulsory license. Failure to work within 4 years, or if license is issued after 5 years, or working is discontinued for 2 consecutive years, patent considered lapsed.

Brunei

Patents granted as confirmation of and coterminous with U.K., Malyasia, or Singapore patents, to be applied for within 3 years of grant in latter countries. No novelty examination, working, or compulsory licensing provisions.

Bulgaria

Invention patents valid 15 years after application; law also provides for inventors' certificates. Some medical and biological inventions eligible only for certificates. Prior publication or public knowledge anywhere prejudicial. Novelty examination. No opposition. Working and licensing provision; 3 years after grant or 4 years after application. Importation may qualify.

Burma

No patent law; Indian patents valid.

Burundi

Invention patents valid for 20 years from application filing date; importation patents valid to 20 years. Public use anywhere prejudicial. No novelty examination or opposition. Patent must be worked within 2 years, otherwise can be cancelled. No compulsory licensing provision.

Canada

Patents valid 17 years after grant. Chemical manufacturing processes patentable. Prior knowledge, use, patent, or description anywhere, or public use or sale in Canada more than 2 years prior to Canadian application prejudicial. Canadian application must be filed either before grant of first foreign patent or within 12 months of first foreign application. Novelty examination. Opposition period not provided, but protest may be filed. Compulsory licensing may be ordered by Patent Commissioner 3 years after grant, and if licenses are insufficient, patent may be revoked. Patent registration should be marked on product.

Chile

Invention patents valid up to 15 years after grant, including renewals. If invention is patented abroad, patent is coterminous with original foreign patent. Prior public knowledge, use, sale, or publication anywhere, or importation into Chile prejudicial. Foreign patent not prejudicial, if invention not commercially known in Chile. Novelty examination. One month opposition period. No working or compulsory licensing provisions. Patent registration must be marked on product.

China, People's Republic Of

Patent law promulgated March 12, 1984 is effective April 1, 1985. The new law shall grant patent protection for 15 years from date of filing application. Foreigners may file applications based on reciprocity or relevant treaty or convention. Twelve month priority period recognized from date of foreign filing. Pharmaceutical products and substances obtained by means of chemical process are excluded from patentability. However, the processes themselves may be patented. Working of patents required from 3 years of issuance. Compulsory licenses available in case of non-working. Preliminary and novelty examinations conducted. Novelty requires idea to be new, not only not known in China, but in the world.

Columbia

Invention patents valid from 5 years after grant, renewable for additional 5 years if subject matter sufficiently worked in Columbia. Use or publication anywhere prejudicial to novelty. Foreign patents must be filed prior to 1 year following application in country of origin. Formalities and novelty examination conducted. Pharmaceutical products, foodstuffs, and beverages not patentable; however, processes may be patented. Opposition 90 days following publication of application. Working required within 3 years following issuance of patent. Working not to be interrupted more than 1 year. Compulsory licenses may be granted if patent not adequately worked in Columbia.

Costa Rica

Invention patents valid 20 years after grant; confirmation patents coterminous with basic patent up to 20 years. Prior public knowledge or use in Costa Rica prejudicial for confirmation patents, anywhere for other patents. Novelty examination. Opposition period 30 days. Working required 2 years after grant, not to be interrupted for any 3 year period; no compulsory licensing provision.

Cyprus

Patents only obtainable as confirmation of U.K. patents. Request must be made within 3 years of U.K. patent grant date.

Czechoslovakia

Invention and dependent patents valid 15 years after application. Chemical and medicinal manufacturing processes patentable; certain medical treatments and biological inventions eligible only for inventor's certificates. Prior public knowledge anywhere via publication, patent description, or display prejudicial. Novelty examination. Opposition period 3 months. No working requirement, but use of patent in public interest may be ordered 4 years after filing or 3 years after grant, whichever is later. Compulsory licenses may be granted.

Denmark

Invention and dependent patents valid 17 years after application, addition patents coterminous with basic patent. Prior public description or public use anywhere prejudicial. Novelty examination. Opposition period 3 months. Compulsory licensing possible 3 years after grant or 4 years from application filing date, if inadequately worked.

Dominican Republic

Invention and revalidation patent valid up to 15 years

after grant, including renewals, medicines and chemicals patentable if approved by medical board. Public knowledge or use anywhere prejudicial. No novelty examination. Foreign patenting or importation no bar. No opposition provision. Working required 5 years after application, not to be interrupted for any 3 year period. No compulsory licensing provision.

Ecuador

Invention patents valid up to 12 years after grant. Importation patents, valid up to 12 years, apply regionally or over whole country. Revalidation patent based on, and coterminous with, patent owned in U.S. or certain latin American countries. Prior existence or public knowledge in Ecuador prejudicial. Opposition period 90 days. If working 2 years after grant, if inadequately worked, or if working interrupted for any 2 year period, patent may lapse. Compulsory licensing possible if patent not worked.

Egypt

Invention patents valid 15 years after application, renewable 5 years; food and drug process patents valid 10 years. Public use or publication in Egypt prejudicial. Novelty examination. Opposition period 2 months. Compulsory licensing possible 3 years after grant if inadequately worked, if working interrupted for any 2 year period, of if needed to work another invention. Revocation possible 2 years after first compulsory license.

El Salvador

Invention patents valid up to 15 years after grant, renewable 5 years in exceptional cases. Prior publication (except in foreign patent documents) or public use anywhere prejudicial. No novelty examination. Patent can be applied for based on foreign patent, if no other publication occurred. Opposition period 90 days. Patent must be worked within 3 years of grant and working not interrupted for more than 3 years; otherwise subject to compulsory licensing. Patent markings on products are compulsory.

Ethiopia

No patent law. Publishing cautionary notices in local press and informing government of existence of foreign patent may afford some protection.

Fiji

Patents valid for 14 years from the grant, or until corresponding first foreign patent expires. Confirmation patents coterminous with U.K. patents also granted. No novelty examination. Use or publicity in Fiji prejudicial. No working or compulsory licensing provisions.

Finland

Invention patents valid 17 years after application. Prior publication in any form anywhere or public disclosure prejudicial. Novelty examination. Opposition period 3 months. Compulsory licensing or revocation possible 3 years after grant, if inadequately worked, or if needed to work another patent.

France

Invention patents valid 20 years from filing date. Law also provides for "certificates of utility," issued for 6 years. Public knowledge anywhere, including publication of a corresponding patent in an official journal, prejudicial. Novelty examination for invention patent, but not for utility certificate applications. Compulsory licensing possible 3 years after grant or 4 years after patent filing date, if inadequately worked, or if working discontinued for any 3 year period. Special legislation applies to patenting of pharmaceutical cases. If no request made during that period, application converted to one for "certificate of utility." No opposition provision, but application laid open for 18 months to permit public comment. Patents applied for before new law effective date (January 1, 1969) remain subject to former law.

Gambia

Patents only obtainable as confirmation of U.K. patents. Request must be made within 3 years of U.K. patent grant date.

German Democratic Republic

Exclusive, non-exclusive (economic), and addition patents granted. Patents valid for 18 years from application date, except for patents of addition which are valid for the remaining term of the main patent. Publication, use in East Germany, or description in a printed publication anywhere prejudicial to novelty. Exclusive patents vest ownership rights on registrant, non-exclusive patents vest rights on registrant and also on any third parties authorized by the patent office. Patents must be worked in East Germany. No provision for compulsory licensing. However, based on public need after indemnification, government may order restriction or revocation of the patent. Patent Court of Leipzig has jurisdiction in infringement cases.

Germany, Federal Republic Of

Invention patents valid 18 years after application if filed before January 1, 1978; 20 years if filed after December 2, 1977. Prior public use in Germany or printed descriptions anywhere (including patent applications and registered utility models) prejudicial, if filed before January 1, 1978; absolute novelty; not made available anywhere, if filed after December 31, 1977. All applications given preliminary screening. Applications opened for public inspection for 18 months after filing. Applicant can postpone full novelty examination request for 7 years; if no request made by that time, application lapses. Opposition period 3 months. Compulsory licensing possible at any time; revocation, adequately worked. Patenting in U.S. constitutes working in Germany.

Ghana

Patents only obtainable as confirmation of U.K. patents, except those for pharmaceutical. Request must be made within 3 years of U.K. patent grant date.

Greece

Patents valid 15 years after application. No novelty examination. No opposition provisions. Revocation possible 3 years after grant if inadequately worked; compulsory licensing possible if needed to work another patent. Advertised offers of licensing may be considered working. Owners of corresponding U.S. patents exempted from working requirement.

Guatemala

Invention patents valid 15 years after grant; importation patents coterminous with basic patent up to 15 years. Inventions to be patentable must meet positive criteria listed in the patent law; prior public knowledge in Guatemala prejudicial for importation patents, anywhere for other patents. Novelty examination. Opposition may be filed within 40 days after first publication. Compulsory licensing possible 1 year after grant if inadequately worked or if working interrupted for any 3 month period.

Guyana

Invention patents valid 16 years after application. Publication, working, sale, or use in Guyana prejudicial. Novelty examination. Opposition period 2 months. Compulsory licensing or revocation possible after 3 years if inadequately worked. Patent registration should be marked on product. Confirmation patents based on U.K. registration also issued.

Haiti

Invention and revalidation patents valid up to 20 years with grant, including renewals; addition patents coterminous with basic patent. Public use or publication anywhere more than 1 year prior to application prejudicial. No novelty examination, opposition, compulsory licensing, or working provisions.

Honduras

Invention patents valid up to 20 years after grant for

foreigners, granted only for life of basic foreign patent. Prior publication or use in Honduras prejudicial for import patents, anywhere for other patents. Novelty examination. Opposition period 90 days. If patent is not worked within 1 year after grant, patent lapses. Patent notice marking necessary to maintain infringement actions.

Hong Kong

Patents only obtainable as confirmation of patents granted in the U.K. Application for protection must be filed within 5 years of U.K. patent issue date. Hong Kong patent expires with corresponding U.K. patent.

Hungary

Patents valid 20 years from application filing date. Application first examined for formalities and published, if accepted. Applicant or others then have 4 years to request complete novelty examination, otherwise application considered abandoned. Prior use or publication anywhere prejudicial. Opposition period 3 months after full examination. Compulsory licensing if patent not worked within 4 years of filing or 3 years from grant, whichever is later.

Iceland

Invention patents valid 15 years after grant. Prior publication anywhere, including use or display in Iceland, prejudicial. Novelty examination. Opposition period 12 weeks. Compulsory licensing possible 5 years after grant if inadequately worked or 3 years after grant if needed to work another patent.

India

Invention patents valid 14 years after application, other than for foods and drugs. Food and drug process patents valid 7 years from filing date. Prior public knowledge or use in India prejudicial. Novelty examination. Opposition period 4 months. Compulsory licensing possible 3 years after grant for inadequate working. All patented articles must be marked with number and year of patent. Reciprocal priority rights granted on basis of applications filed in certain Commonwealth countries. Not member of Paris Union. Patents granted before April 20, 1972 exist 20 years from filing.

Indonesia

Pending passage of patent legislation, applications may be filed with the Indonesia Justice Ministry. Although applications will not be acted on until a patent law is passed, they will reportedly be considered regular applications.

Iran

Invention patents valid up to 20 years after application; addition or improvement patents coterminous with foreign patents. Except for import patents, prior publication in official publications or journals anywhere prejudicial. No novelty examination. Opposition period not specified. Revocation 5 years after grant if inadequately worked. No compulsory licensing provision.

Iraq

Invention patents valid 15 years from date of application, patents of importation for unexpired term of their foreign basic patents up to 15 years. Prior public knowledge or use anywhere prejudicial. No opposition provision. Patent must be worked within 3 years and not discontinued for 2 years, otherwise subject to compulsory license or revocation.

Ireland

Invention patents valid 16 years, renewable under exceptional conditions up to 10 years when inadequately remunerated. Prior public use or knowledge anywhere prejudicial. Novelty examination. Opposition period 3 months. Compulsory licensing possible 3 years after grant or 4 years after application date, if inadequately worked. Revocation possible 2 years after first compulsory license. Importation does not constitute working. Patent registration should be marked on product.

Israel

Invention patents valid 20 years after application prior publication, use, or sale anywhere prejudicial. Novelty examination. Opposition period 3 months. Compulsory licensing possible 3 years after grant or 4 years after application date,. if adequately worked; revocation possible 2 years after first compulsory license.

Italy

Invention patents valid 20 years after application. Prior pubic knowledge anywhere prejudicial. No novelty examination or opposition provisions. Compulsory licensing possible if inadequately worked 3 years after grant or 4 years after application, or if working interrupted for any 3 year period. Exhibition, but not importation, may constitute working.

Jamaica

Invention patents valid 14 years after application, renewable 7 years; confirmation patents based on foreign patents coterminous with original. Prior publication or public use in Jamaica prejudicial, except for confirmation patents. No novelty examination. No opposition provision. No working or compulsory licensing provisions.

Japan

Invention patents valid 15 years from date application published; cannot exceed 20 years application filing date. Application open to public inspection 18 months from filing date. Examination can be postponed 7 years at applicant's request; if no request for examination made at that time, application lapses. Publication anywhere or public knowledge or use in Japan prejudicial. Novelty examination. Opposition period 2 months. Compulsory licensing possible 3 years after registration or 4 years after filing, if inadequately worked or if needed to work another patent.

Jordan

Invention patents valid 16 years after application. Prior

publication, use, or sale in Jordan prejudicial. Novelty examination. Opposition period 2 months. Compulsory licensing or revocation possible 3 years after grant, if inadequately worked.

Kenya

Patents only obtainable as confirmation of U.K. patents. Request must be made within 3 years of U.K. patent grant date.

Korea, Republic Of

Invention patents valid 12 years after grant or 15 years after application, whichever is less. Public knowledge or use in Korea, or appearance in publications distributed in Korea prejudicial. Novelty examination. Opposition period 2 months. Compulsory licensing or cancellation possible 3 years after grant if inadequately worked or if working interrupted for any 3 year period. Patent registration must be marked on product.

Kuwait

Invention patents granted for 15 years from application date, extendable for 5 additional years. Patents of addition granted for remaining term of main patent. Patents for special processes or means for foodstuffs, medicines, or pharmaceutical preparations granted for 10 years. Novelty: invention not published or publicly known in Kuwait for 20 years preceding application. No novelty examination, only for compliance with formal requirements. Working within 3 years following grant not to be interrupted for more than 2 years, otherwise compulsory licenses may be issued.

Lebanon

Invention patents valid 15 years after application. Prior publication anywhere prejudicial. No novelty examination. Opposition provision. Working required 2 years after grant. For nationals of Paris Union countries, however, allowance period is 3 years. No compulsory licensing provisions. Direct offer to license party capable of working the invention may constitute working, but importation does not.

Lesotho

Confirmation patents based on Republic of South Africa patents; issued for remaining term of corresponding patent in latter country.   U.K. patent automatically protected for its duration; no local registration or confirmation required. Marking of item to indicate patent desirable.

Liberia

Invention patents valid 20 years after grant.   Prior public knowledge, publication, or use in Liberia prejudicial. No specific exclusion from patentability.   No novelty examination.   No opposition provision.   Working required 3 years after grant.

Libya

Invention patents valid 15 years after application, renewable 5 years.   Processes for making foodstuffs, medicines, or pharmaceutical preparations patentable for 10 years.   Public use or publication in Libya prejudicial.   No novelty examination.   Opposition period 2 months.   Working required in Libya or country of origin within 3 years; 2 years extension possible.   Compulsory licensing may be ordered at any time.

Liechtenstein

Swiss patents automatically valid without any required formalities.

Luxembourg

Invention patents valid 20 years after application.   Prior public knowledge or use anywhere prejudicial.   No novelty examination or opposition provision.   Compulsory licensing possible after 3 years; non-working for 3 years can also result in revocation.

Malawi

Invention patents valid 16 years from application filing date.   Prior public use or knowledge in Malawi or in printed publications anywhere prejudicial.   No novelty examination.

Opposition period 3 months. Compulsory license can be ordered, if patent not sufficiently worked within 3 years.

Malaysia

Federation consisting of former areas of Malay, Sabah, and Sarawak; each still has separate patent law. In Malaya and Sabah, U.K. patent is applicable; it must be registered separately within 3 years of U.K. grant, in each territory to be in force; remains in force for the duration of U.K. registration. In Sarawak, application may be filed anytime, based on patent in the U.K., Singapore, or Malaya.

Malta

Invention patents valid 14 years from application, renewable 7 years. Publicity anywhere prejudicial. No novelty examination. Opposition period 2 months. Compulsory licensing possible, if not worked in 3 years.

Mauritius

Invention patents valid 14 years from application, renewable for like period. Publicity or use in Mauritius prejudicial. No novelty examination. Opposition period 1 month. No working or compulsory licensing.

Mexico

Invention protected by patents or certificates of invention at applicant's option. Patents grant exclusive rights to owner to use or authorize others to use invention. Inventor's certificates are available for non-patentable products (chemicals, nuclear inventions, and antipollutant devices)owner of inventors' certificate is under obligation to grant non-exclusive license to any third party wanting to use it, but retains right to continue working the invention. Decisions on payment left to parties, but government may intervene if no decision reached. Patents, patents of improvement, and certificates granted for 10 years from this issue date. Recent prior publication or use anywhere prejudicial. No opposition provisions.

Working of invention is required within 3 years from issue of patent. Patent will lapse, if working not commenced within 4 years from issue and no compulsory license granted. Importation of patent product does not constitute working. Patent subject to compulsory license if (1) not worked within 3 years of issue, (2) working suspended more than 6 months, (3) working fails to meet national demand, and (4) export markets exist which are not being supplied by working of the patent. Government determines duration, scope and royalties payable under such patent license.

Monaco

Invention patents valid 20 years after application. Publicity or use anywhere prejudicial. No novelty examination or opposition. Working required within 3 years after grant, otherwise subject to compulsory licensing.

Morocco

Although country consists of former French Morocco, Tangier zone, and Spanish zone, no unified patent law yet exists. Separate application for the former French Morocco must be filled with the Industrial Property Office at Casablanca, and for Tangier, with industrial Property Bureau in that area. Situation in ex-Spanish zone unclear.

For Morocco, invention patents valid 20 years after application. Prior public knowledge anywhere prejudicial. No opposition provision. Working required 3 years after grant for nationals of Paris Union countries, 3 years after application for others; working not to be interrupted for any 3 year period. Reasonable offer to license or sell may constitute working; importation may be prejudicial. No compulsory licensing provision.

For Tangier zone, invention patents valid 20 years and importation patents 10 years from application. Public use or publication anywhere prejudicial for basic invention patents. No novelty examination or opposition provision. Working required within 3 years, otherwise subject to compulsory licensing.

Namibia (Southwest Africa)

Patent matters administered by South African Patent Office. Invention patents valid 14 years from application filing date, renewable 7-14 year periods. No novelty examination. Prior use in Namibia or abroad more than 2 years before filing prejudicial. No opposition. Compulsory license possible if not worked within 2 years.

Nauru

Patents granted as confirmation of and coterminous with Australian patents. Must be applied for within 3 years of latter's grant date. No working or compulsory licensing provisions.

Nepal

Invention patents valid 15 years after grant, extendable for two additional 15 year periods. Applications screened for novelty regarding known use in Nepal; rejected if criteria not met.

Netherlands

Invention patents granted after January 1, 1978 valid 20 years from effective filing date. Patents granted before January 1, 1964, or based on applications pending that date, valid 18 years from date of grant. Patents applied for after January 1, 1964 and granted before January 1, 1978, valid 20 years from filing date or 10 years from grant, whichever is longer. Prior publication, public knowledge, or use anywhere prejudicial. Novelty examination. Applicant has 7 years after filing to request complete examination; if no request made by then, application lapses. Opposition period 4 months. Compulsory licensing possible 3 years after grant if inadequately worked; at any time if needed to work another patent. Patent registration should be marked on product.

Netherlands law applies to Netherlands Antilles and Surinam.

New Zealand

Invention patents valid 16 years after application; renewable up to 10 years. Recent prior publication in patent specification anywhere, or publication or use in New Zealand prejudicial. Novelty examination. Opposition period 3 months. Working required within 3 years after grant.

Subject to compulsory licensing at any time for foods or medicines; revocation after 2 years, if licensing unsatisfactory.

Nicaragua

Invention patents valid 10 years after grant. Public knowledge in Nicaragua prejudicial. Processes patentable, but not products. Examination. Opposition period 30 days. Working required 1 year after grant, not to be interrupted for any 1 year period. Sworn affidavit and advertised offer of sale or license constitutes working. No compulsory licensing provision. Patent registration should be marked on product.

Nigeria

Patents granted under former law as confirmation of U.K. patents continue until term expires. New law effective December 1, 1971. Invention patents valid 20 years after application. Use or publication anywhere prejudicial. No novelty examination or opposition provisions. Compulsory licensing possible 3 years from grant or 4 years from application, if inadequately worked.

Norway

Since 1980, invention patents are valid for 20 years from application date. Patents granted between January 1, 1968 and January 1, 1980 have a 20 year validity term. Patents granted before 1968 have a 17 year validity. Prior publication and use anywhere prejudicial. Novelty and formal examination. Opposition period 3 months. Compulsory licensing 3 years after grant or 4 years after application if inadequately worked, if working interrupted, or if needed to work another patent.

Oman, Qatar, United Arab Emirates (Abu Dhabi, Dubai)

No patent laws. Cautionary notice in certain Lebanese newspapers circulating in these states.

Pakistan

Invention patents valid 16 years after application. Prior public knowledge or use in Pakistan prejudicial. Novelty examination. Opposition period 4 months. Working required within 4 years of grant, otherwise compulsory licensing possible. Patent registration should be marked on product. Not member of Paris Union, but reciprocal priority rights granted to applications filed in certain Commonwealth countries.

Panama

Invention patents valid up to 20 years. Revalidation patents coterminous with foreign patents up to 15 years. Prior public knowledge in Panama prejudicial. No novelty examination. Opposition period 90 days. Patents may lapse if not worked when one third of term has passed; working not required for revalidation patents. No compulsory licensing provision.

Paraguay

Invention patents valid 15 years after application; confirmation patents coterminous with basic patents up to 15 years. Publication of foreign patent 1 year prior to Paraguayan application, or prior working or public disclosure in Paraguay prejudicial. No novelty examination. No opposition provision. Compulsory licensing possible if not worked for any 3 year period.

Peru

Invention patents valid 10 years from grant. Provisional patents based on foreign patents, valid 1 year only for person domiciled in Peru. Renewed foreign patents valid for unexpired term of foreign patent, but not in excess of 10 years duration. Formal and novelty examination. Prior use in Peru or publication anywhere and not later than 1 year from filing date of first foreign application, prejudicial. Opposition period 30 days after application last published.

Working required within 2 years of grant; period extendable 2 years. Compulsory licensing possible if patentee importing product, local working insufficient, or needed to work another patent.

Poland

Regular patents valid 15 years from application filing date. Applications published 18 months after examination for formalities. Full examination must be requested within 6 months of publication. Applicant can request provisional examination and grant of provisional patent for 5 year duration. Has 4 years after application date to request full examination and conversion to regular patent, if provisional patent sought first. Law also embodies inventors' certificate system, with state assuming ownership of inventions thereunder and granting awards based on invention's use.

Portugal

Invention patents valid 15 years after application. Prior publication anywhere or public use in Portugal prejudicial. Novelty examination in case of opposition. Opposition period 3 months. Compulsory licensing possible 3 years after grant if inadequately worked, if working interrupted for any 3 year period, or if needed to work another patent.

Romania

Invention patents valid 15 years after application; patents of addition for period of basic patent, but no less than 10 years. Inventors' certificates provided for in law. Filing acceptable if invention not previously filed or patented in Romania or publicly revealed anywhere. No opposition provision. Working required 4 years after application or 3 years after grant; otherwise subject to compulsory licensing.

Rwanda

Invention patents valid 20 years after application. Importation patents coterminous with corresponding foreign patents, not to exceed 20 years. Public use in Rwanda or publication anywhere prejudicial. No novelty examination or opposition. Patent must be worked within 2 years from date worked abroad, otherwise can be cancelled. No compulsory licensing.

Ryukyu Islands (Okinawa)

Reverted to Japan May 15, 1972. Japanese patent law extends to this area.

San Marino

Industrial property rights obtained in Italy applicable.

Saudi Arabia

No patent law. Can publish cautionary ownership notice in local *Official Gazette* for such rights this may offer in seeking court action against infringers.

Sierra Leone

Patents only obtainable as confirmation of U.K. patents. Request must be made within 3 years of U.K. patent grant date.

Singapore

Patents only obtainable as confirmation of U.K. patents. Request must be made within 3 years of U.K. patent grant date.

Somalia

Patents granted for 15 years from filing date. Prior knowledge of inventions anywhere prejudicial. Patent must be worked within 3 years of grant and working not interrupted for 3 consecutive years.

South Africa

Patents valid 20 years after application. Prior public knowledge, use, or working in South Africa, or publication anywhere prejudicial. No novelty examination. Opposition period 3 months. Compulsory licensing possible 3 years after grant or 4 years after application, whichever is later, if inadequately worked, if needed to work another patent, or for foods, plants, or medicines. Importation does not constitute working. Patent registration should be marked on product.

Spain

Invention patents valid 20 years after grant. Patents of importation valid 10 years and may be applied for by anyone. Recent public knowledge or working in Spain prejudicial for import patents, anywhere for other patents. No novelty examination. Compulsory licensing required 3 years after grant if not worked. Importation patents must be worked annually to remain in force.

Sri Lanka

Invention patents valid for 15 years after grant. Prior public knowledge anywhere via publication, description, or use prejudicial. Examination for compliance with formalities and novelty. Opposition not provided for. Working and compulsory licensing not provided for in new Code (August 8, 1979)However, the Code's transitory provisions may require compulsory licenses or revocation of the patent if not worked properly within 3 years following application or priority date.

Sudan

No patents issued. Can publish cautionary ownership notice of foreign patent in local *Official Gazette* for such rights this may offer in seeking court action against infringers.

Swaziland

Confirmation patents based on prior registration in South Africa. U.K. patents automatically in force.

Sweden

Invention patents valid 20 years after application. Prior publication or public use anywhere prejudicial. Novelty examination. Opposition period 3 months. Compulsory licensing possible 3 years after grant or 4 years after application, if inadequately worked, or if needed to supplement an earlier patent.

Switzerland

Invention patents valid 20 years after application. Prior

public knowledge in Switzerland or public disclosure anywhere prejudicial.   Textile and timepiece inventions subject to novelty examination.   Opposition period 3 months. Compulsory licensing possible 3 years after grant, if inadequately worked, or if needed to work another patent. Working in U.S. satisfies working requirement.

Syria

Invention patents valid 15 years after application.   Prior public knowledge anywhere prejudicial.   No novelty examination.   No opposition provisions.   Working required 2 years after grant.   No compulsory licensing provisions.

Taiwan

Invention patents valid 15 years after publication; addition patents coterminous with basic patent.   Prior publication or public use prejudicial.   Novelty examination. Opposition period 3 months.   Compulsory licensing or revocation possible, if inadequately worked within 3 years from grant.   Patent registration should be marked on product.

Tanzania

Consists of former Tanganyika and Zanzibar areas; now joined as an independent state; former separate laws for each are still in effect.   In former Tanganyika, confirmation of U.K. patents only for term of U.K. patent; application for confirmation patent must be made within 3 years of U.K. patent grant.   In former Zanzibar, same situation prevails, except that registration in that area may be invalidated by manufacture, use, and sale of invention subject matter before priority date of U.K. patent.

Thailand

New patent law, published March 16, 1979, entered into force September 12, 1979.   Invention patents valid for 15 years from application filing date.   Public use in Thailand or disclosure of invention anywhere prejudicial to novelty. Foreign applicants may apply only if their countries permit Thai nationals to file applications.   Foreign applications may be ordered to be accompanied by an examination conducted by a foreign government or other appropriate organizations dealing with patents. If the application meets certain basic formalities and substantive criteria, it is published. The applicant must then submit a request for novelty examination within 5 years after publication, otherwise the application is considered abandoned. Third parties may file opposition within 180 days of the application's publication. Compulsory licensing possible 3 years after patent issues, if not properly worked. After 6 years of non-working, patent may be revoked by government. Product design patents issued for 7 years from application date. Imports of patented products are prohibited, except as specifically permitted, on request, by the government.

Trinidad and Tobago

Invention patents valid 14 years after grant, renewable for 7 years. Prior public use in Trinidad or Tobago prejudicial. No novelty examination or opposition. Compulsory licensing possible.

Tunisia

Invention patents valid up to 20 years after application. Prior publication, public knowledge, or public use anywhere prejudicial. No novelty examination. Opposition period 2 months. Working required 3 years after grant (2 years for non-Paris Union nationals), not to be interrupted for any 2 year period. Importation not considered working; could invalidate patent. No compulsory licensing provision.

Turkey

Patents valid up to 15 years after application. Prior publication anywhere prejudicial. Novelty examination. No opposition provision. Compulsory licensing possible 3 years after grant if inadequately worked or if working interrupted for any 2 year period.

Uganda

Patents only obtainable as confirmation of U.K. patents. Requests must be made within 3 years of U.K. patent grant date.

United Arab Emirates

See Oman, Qatar.

United Kingdom

For patents filed before June 1, 1967, duration is 16 years from filing date. Can be extended for 10 years on applications filed before June 1, 1978, and up to 4 years on applications filed after June 1, 1978. For patents dated after June 1, 1967, duration 20 years from filing date on original patent. For applications filed after June 1, 1978, duration is 20 years from filing date. Prior public disclosure in the U.K. or abroad prejudicial. Novelty examination. Compulsory licensing possible 3 years after grant, if inadequately worked, for patents on applications filed before June 1, 1978.

Certain newly independent countries in the British Commonwealth have already established or are in the process of developing their own national patent codes. In the meantime, some such countries continue to use pre-independence procedures and facilities in providing patent protection within their respective territories. These countries, as well as those of the British Commonwealth that now have separate patent laws, are covered separately and alphabetically. Generally, in such countries the patent protection available is by registered confirmation of a U.K. patent which must take place within 3 years of the original U.K. grant.

Uruguay

Invention and related improvement patents valid 15 years after grant; revalidation patents granted for unexpired term of foreign patents, but not to exceed 15 years; must be applied for within 3 years of basic patent. Prior public knowledge anywhere prejudicial. Novelty examination. Opposition period 20 days. Compulsory licensing possible 3 years after grant if not worked. Importation or nominal working does not constitute working.

U.S.S.R.

Soviet law provides for granting of either patents or

inventors' certificates for new inventions. Patents are granted for 15 years' duration after application. Inventors' certificates have limited duration. Full examination is made for novelty and usefulness, based on prior Soviet and foreign patents and publications and prior Soviet inventors' certifications. No opposition provision. No working provision. Compulsory licensing possible.

Venezuela

Invention and improvement patents issued for 5 or 10 years after grant at owner's request. Revalidation patents (based on prior foreign filing) coterminous with basic patents issued up to 10 years. Patents of introduction available to non-owner of foreign patent are granted for a 5 year term, but do not protect against imports. Publication in Venezuela or public knowledge anywhere prejudicial for invention or improvement patents. Prior public knowledge in Venezuela prejudicial for introduction and improvement patents. Opposition period 60 days. Working for all patents must be effected 2 years after grant. Thereafter, working must continue for 2 consecutive years for 10 year patent and 1 year for 5 year patent.

Yemen Arab Republic

Patent applications accepted under Law No. 45 of 1976, although no patent yet issued to foreigners.

Yugoslavia

Patents valid 15 years after application publication. Publication or description anywhere, or sale, use, or display in Yugoslavia prejudicial. Novelty examination. Opposition period 3 months. Compulsory licensing possible 3 years after grant if inadequately worked.

Section 2

Export Regulations and Technology Transfer

One of the primary emphases in this book has been the marketing of technology, or out-licensing, to international markets. Indeed, this has traditionally been one of the major elements of technology transfer, and this trend will doubtless continue as the need for technology and the capability of using it grows throughout the world.

In the main body of the book, we addressed commercial aspects and activities of technology transfer in international markets. This coverage included chapters on commercial activities such as market research, product packaging, sales and negotiation, and licensing agreement development. In the course of that coverage, we noted, but did not highlight, another aspect of technology transfer, that of government regulations governing exports. including technology itself. Although not strictly a business/commercial issue of and by themselves, the export regulations can impact nearly all other activities relating to the business of technology transfer.

Consider, for example, some of the questions that have actually arisen relating to the application of Export Administration Regulations to the various elements of technology transfer:

o   Does the law permit us to supply

> preliminary licensing information in response to this inquiry from Eastern Europe?

o   At what point in the marketing effort will we require an export license?

o   Are there restrictions on what we can bring with us on a sales call to a Bulgarian client's office?

o   Are there certain countries we should eliminate from our technology marketing plan because of export restrictions?

o   Will the export regulations affect our planned participation in an international trade show?

Clearly, the answers to such questions can have a significant impact on the technology marketing effort. In addition, these answers are obviously affected by the nature of the technology, the identity of the potential customer, and, above all, the content of the regulations. Accordingly, we shall present a brief overview of the export licensing process, followed by those sections of the export regulations that are applicable to technology transfer transactions.

Export controls have been established by the U.S. and other countries because of national security or foreign policy considerations, and also to deal with domestic products which may be in short supply. One usually thinks of export regulations as applying only to tangible "high-tech" items with obvious military and strategic significance. However, the Export Administration Regulations apply also to intangible intellectual property, such as technology and know-how.

There are two generic types of export licenses; general and validated. A general license involves a broad grant covering certain categories of products to most destinations. Although there are about twenty types of general export licenses, the category that usually applies to technology transfer transactions is the *General License for Technical Data under Restriction (GTDR)*. A general license does not require a specific application, but may require additional documentation, such as a "Destination Control Statement" in certain instances.

In order to determine whether a validated license is required, several criteria apply.   First, the schedule of "Country Groups" in the export regulations must be checked against the destination of the specific technology.   Second, the nature of the product must be reviewed against the "Commodity Control List" and special restrictions to determine the need for a validation license.   Most intangible technology products are covered under the general category of "unpublished" technical data, which refers to technical information not available to the public, or know-how that the owner will not release without some form of compensation. Here again, in addition to the application for validated export license, where required, care must be taken to establish whether or not additional documentation will be required from the proposed recipient of the technology or his government.

For more specific information or guidance, one should contact:

> Export Assistance Staff
> Room 1099D
> U.S. Department of Commerce
> Washington, DC 20230
> (203) 377-4811

Forms needed to comply with the regulation are available from:

> U.S. Department of Commerce
> Operations Support Staff
> P. O. Box 273
> Washington, DC 20004
> ATTN: Forms

Finally, reference is made to the following pages, wherein the portions of the Export Administration Regulations applicable to technology transfer are reproduced. More specifically, there are included the following excerpts from the regulations themselves:

o   Introduction to the Export Administration Regulations

o   Country Groups

o   Part 379—Technical Data Export Licenses

o   Part 385—Special Country Policies

o   Application for Export License

o   Export License

INTRODUCTION TO THE EXPORT ADMINISTRATION REGULATIONS

The Export Administration Regulations are issued by the Department of Commerce to enforce the Export Administration Act of 1979. In passing this legislation, the Congress listed three general policy guidelines for the use of export controls.

First, controls should be used on exports "which would make a significant contribution to the military potential of any other country or combination of countries which would prove detrimental to the national security of the United States."

Second, controls should be used "where necessary to further significantly the foreign policy of the United States or to fulfill its declared international obligations."

Third, controls should be used "where necessary to protect the domestic economy from the excessive drain of scarce materials and to reduce the serious inflationary impact of foreign demand."

This Introduction is a simple guide to the Export Administration Regulations. It is intended to help you comply with the Regulations. It is informational only and does not substitute for or modify the official rules. It does, however, contain a summary of each Part of the Regulations and refers you specifically to those Parts you most likely will need to review. On the opposite page is a chart that displays the essential steps to follow once you receive an export order.

GENERAL VS. VALIDATED LICENSE AUTHORIZATION

The first step is to find out whether your shipment requires a validated license or whether you may export under one of the General License authorizations that are defined in Part 371 of the Regulations.

You probably will have to apply for a validated export license if you are exporting—

A "strategic" commodity to any destination; or, in a few cases, only to a destination to which exports are restricted for national security purposes, e.g. certain communist destinations;

A "short supply" commodity to any destination;

Any other commodity to a destination for which there are foreign policy concerns; or

"Unpublished" technical data to certain destinations.

Generally speaking, a "strategic" commodity is one that we believe is capable of contributing signifi-

cantly to the design, manufacture, or utilization of military hardware. The fact that the commodity might also have peaceful uses does not remove the "strategic" label.

A "short supply" commodity is one that we have found to be in short supply in the United States *and* which is wanted abroad. If permitted to be exported without restriction, there would be an excessive drain on U.S. supplies and a serious inflationary impact on the U.S. economy.

The term "unpublished" technical data means technical information, generally related to the design, production, or use of a product, that is not available to the public. It is not described in detail in books, magazines, or pamphlets nor is it taught in colleges or universities. It is know-how that a person will not release without charging for it.

You will have to refer to the Commodity Control List, Supplement No. 1 to § 399.1 of the Regulations, to find out if the commodity you intend to export is classified as a "strategic" or a "short supply" commodity or whether it falls in the "any other" category.

First, check the code letter that follows the Export Control Commodity Number for the entry in which your commodity is located. If the code letter is "A" or "B," the commodity is under validated license control to all destinations, generally excepting Canada. If the code letter is "C," "D," or "E" the commodity is under validated license control to certain communist destinations, but not to Free World destinations.

Second, check the paragraph titled "Reason for Control." Commodities are controlled for reasons of national security, short supply, foreign policy, nuclear non-proliferation and crime control (foreign policy).

If you find your commodity is not controlled for "strategic" or "short supply" reasons, you will need to file an application *only* if the country of destination is one for which there are foreign policy concerns. At the present time, there is a general embargo on exports to Cuba, Vietnam, North Korea, and Cambodia for foreign policy reasons. There are selective controls, also for foreign policy reasons, for certain types of commodities for export to the Republic of South Africa and Namibia, certain Middle East destinations, and the U.S.S.R., Estonia, Latvia and Lithuania. In addition, foreign policy controls are in effect requiring validated licenses to export commodities useful in crime control and detection to all countries except: Australia, Belgium, Canada, Denmark, France, the Federal Republic of Germany (including West Berlin), Greece, Iceland, Italy, Japan, Luxembourg, The Netherlands, New Zealand, Norway, Portugal, Spain, Turkey, and the United Kingdom.

Read and follow the step-by-step instructions in § 399.1 on how to use the Commodity Control List (CCL). If you still cannot determine whether you must file an application for a license to export to your foreign customer, contact either the Exporter Assistance Staff of the Office of Export Licensing in Washington, D.C. (202) 377-4811, or your nearest Commerce District Office for assistance. The District Offices are listed in the front of the Export Administration Regulations.

GENERAL LICENSE SHIPMENTS

Shipper's Export Declaration

Let's assume you have determined that you may make your shipment under a General License authorization. Your next step is to find out if you need to prepare and file a Shipper's Export Declaration (SED). The SED serves two purposes. First, it provides the Bureau of the Census with data on exports. These data are compiled and published monthly and show what types of commodities are exported to what countries. Second, the SED serves as an export control document. You will be required to show on the SED either the General License symbol (e.g., G-DEST) or, if a validated export license was required and issued, the license number assigned by Commerce. By placing the General License symbol or license number on the SED, you are certifying that the shipment meets the conditions of the general license authorization or of the validated license. This notation assists the Department in its efforts to assure compliance with the Export Administration Regulations. You can find out if an SED is required by reading Part 386 of the Regulations. Note particularly § 386.3 on preparation of the SED.

Destination Control Statement

The final step before exporting under a general license authorization is to find out whether a Destination Control Statement must be placed on your commercial invoice and your bill of lading or air waybill. If the comodity you plan to export is neither a "strategic" nor a "short supply" commodity, the chances are that you will *not* be required to place this statement on your commercial documents. But a statement *is* required if you are

shipping a commodity that may be exported to the Free World under general license but requires a validated license to export to a restricted communist destination and the shipment is valued at more than $1,000. A statement also is required for general license shipments to the Republic of South Africa or Namibia. The details are contained in § 386.6 of the Regulations.

Once the SED is prepared, you should submit it to the exporting carrier or, if the shipment is to be made by mail, to the Postmaster. The your shipment can proceed to its final destination.

Technical Data

So far, we have concentrated on exports of *commodities* under general license authorization. If you plan to export technical data, read Part 379 of the Regulations very carefully. Note that there is a general license authorization for the export of technical data that are generally available to the public, for scientific and educational data, and for certain types of patent applications. These types of data are defined in § 379.3 of the Regulations. Export controls on other types of data, what we call "unpublished" data, vary widely, depending on the country of destination and the commodities to which the data relate. Section 379.4 of the Regulations tells whether you need to apply for an export license, or whether you have to obtain "prior assurances" from your foreign customer. Generally, there is no need to prepare and file an SED before exporting technical data.

APPLICATION FOR A VALIDATED EXPORT LICENSE

Now let's assume that you have reviewed the Commodity Control List and found that you may not export a particular commodity for which you have an order under a General License authorization. You will have to apply for a validated export license.

Advisory Opinions

Because the processing of an export license application costs us money, as a general policy we do not issue what are called "hunting licenses." By this we mean a license to export based merely on a business inquiry or a person's *desire* to obtain foreign business. There must be an "order" from a foreign person or firm for a particular commodity or technical data before we will consider your application. However, the "order" need not be an agreement that can presently be executed or that would become a binding contract upon acceptance. Section 372.6 defines an "order" and spells out the distinction between a firm "order" and a mere business inquiry. We recognize, however, that there are times when an exporter who does not have an order needs to know the likelihood of his being able to obtain a license. For example, negotiations with a foreign buyer may hinge on your prospects of obtaining a license. If you are in this situation, you may write to us, describing the proposed transaction in as full detail as possible at the time and explaining why you need our advice, and we will respond with an advisory opinion. This opinion, while telling you the likelihood of our issuing a license, does not obligate us to issue a license. In short, it is not a commitment. A definitive judgment is rendered by the Department only by the issuance of a formal validated license.

Supporting Documentation

Delays that happen because you have not prepared your application properly, or because you have not submitted the required supporting documentation, are costly both to you and to the Government. Therefore, you should read the instructions in Part 372 and the application instruction sheet carefully before filling out the application. You should also read Part 375 of the Regulations to see whether you need to support your application with an International Import Certificate or a Consignee/Purchaser Statement or other documentation. If you fail to submit any supporting documentation that is required, your application may be returned to you without action. The Forms Supplement to the Regulations contains examples of all forms that may be required to obtain an export license. It also illustrates how to prepare an export license application and consignee/purchaser statement correctly.

Special Commodity Policies

You should also review Part 376 of the Regulations. This Part will tell if there are any special commodity policies or provisions that might pertain to the products that require a validated export license. A quick review of the table of contents for this Part will tell at a glance if you need to read any particular Section of the Part. For example, if you are exporting machine tools or numerical controls, read § 376.11. If you propose to export "short supply" commodities, read Part 377 and be guided by the instructions in that Part.

Status Information

Once your application has been prepared properly and is supported by any document that is required, mail it to The United States Export Administration,

Box 273, U.S. Department of Commerce, Washington, D.C. 20044.

The Department's review of proposed exports of "strategic" commodities to Free World destinations will focus principally on the likelihood that the intended consignee will reexport the product illegally.

If your application is prepared properly and is documented adequately, and if the consignee is located in the Free World and is known to the Department, you should receive your license well within three weeks from the date you mailed your application. Free World applications reviewed for nuclear or human rights purposes, however, may take longer. If your application covers an export to a Free World destination and you haven't received word from the Office of Export Licensing after three weeks, you can request information on the status of the application by calling Exporter Assistance Staff (202) 377-2752. If your application covers an export to any other destination (for example, a restricted communist country), wait at least five weeks before calling.

Shipper's Export Declaration (SED) and Destination Control Statement

Upon receipt of a validated export license, you should prepare the Shipper's Export Declaration (SED). We have already discussed this document. But remember that the validated export license number must be entered on the SED before it is filed with the carrier. You also will be required to place one of three Destination Control Statements on your commercial invoice and bill of lading or air waybill. Refer to § 386.6 for information on these Statements and their uses. When you have the SED prepared properly, submit it to the carrier (or Postmaster) and make your shipment.

Returning the Used License

The final step is to post the shipment on the reverse of the validated export license as explained in § 386.2 of the Regulations, and return the used license to the Office of Export Licensing. However, if you are making more than one shipment of the commodities authorized on your validated export license, hold the license until all shipments have been made or until the license has expired. Remember to post each shipment on the reverse of the license so that it will contain a complete accounting of the use you made of the license. Also, return all *unused* licenses issued to you as soon as they have expired.

SUMMARIES OF THE PARTS IN THE REGULATIONS

PART 368

U.S. Imports

Part 368 covers the two procedures designed to increase the effectiveness of control over international trade in strategic commodities. The *International Import Certificate* (IC), issued by the importing country, certifies that the importer will comply with his country's export control regulations when reexporting the commodities covered by the IC. The *Delivery Verification Certificate* (DV) assures the foreign government that the commodities were not diverted from their intended destination. Part 368 discusses the procedures to follow if you are importing commodities and the exporting country requires an International Import Certificate or a Delivery Verification Certificate. This Part also explains penalties which may be imposed if you violate these Regulations.

Some commodities for which import certificates are required may be controlled by other agencies. Section 368.1(a)(2) explains which commodities are covered by the Export Administration Regulations, and which agencies you should consult for other commodities.

If you are *exporting* commodities from the United States, see Part 375 to find out if you need an International Import Certificate from the importing country.

PART 369

Restrictive Trade Practices or Boycotts

Part 369 implements the antiboycott provisions of the Export Administration Act of 1979. It defines who is covered by the antiboycott rules; what activities are covered; what actions in response to boycott requests are prohibited; what exceptions to the prohibitions are permitted; and what the reporting requirements are. Supplements to Part 369 contain interpretations issued by the Department on boycott-related matters.

PART 370

Export Licensing General Policy and Related Information

Part 370 serves as the foundation for the Export Administration Regulations. It explains the reasons why controls are imposed, and defines the terms which are used throughout the Regulations, although other Parts may include definitions of terms peculiar to those Parts.

This Part discusses the types of shipments for which an export license is required, and explains certain situations for which a license will not be required. However, be certain to read all applicable parts of the Regulations before exporting without a license.

Some commodities are licensed by other U.S. Government agencies. This Part tells which commodities are under another agency's jurisdiction and what office or bureau within that agency to contact for information.

Part 370 also details how applicants can obtain information on various matters and where to get copies of Export Administration forms.

Supplements to this Part provide a color-coded country group map and list the Country Groups referred to throughout the Regulations, as well as list commodities licensed by the Department of State (U.S. Munitions List) and the Nuclear Regulatory Commission (Nuclear Equipment and Material).

PART 371

General Licenses

A general license is a general authorization permitting the export of certain commodities and technical data without the necessity of applying for a license document. Part 371 describes fully the purpose of each general license, restrictions on its use, and special requirements for each. This Part also explains requirements for Shipper's Export Declarations on general license shipments, and lists certain prohibitions against the use of general licenses. (General Licenses GTDA and GTDR pertain to technical data, and are described in Part 379, Technical Data.) NOTE: *The Export Administration Regulations are subject to change and a general license authority for a particular commodity area may be revoked at any time.* There are a number of different general licenses. Their symbols and uses are summarized in the following table:

Export Licensing General Policy
and Related Laws

COUNTRY GROUPS

For export control purposes, foreign countries are separated into seven country groups designated by the symbols "Q", "S", "T", "V", "W", "Y", and "Z". Listed below are the countries included in each country group. Canada is not included in any country group and will be referred to by name throughout the Export Administration Regulations.

Country Group Q

Romania

Country Group S

Libya

Country Group T

North America
  Greenland
  Mexico (including Cozumel and Revilla Gigedo Islands)
  Miquelon and St. Pierre Islands

Central America and Caribbean
  Bahamas
  Barbados
  Belize
  Bermuda
  Costa Rica
  Dominican Republic
  El Salvador
  French West Indies
  Guatemala
  Haiti (including Gonave and Tortuga Islands)
  Honduras (including Bahia and Swan Islands)
  Jamaica
  Leeward and Windward Islands
  Netherlands Antilles
  Nicaragua
  Panama
  Trinidad and Tobago

South America
  Argentina
  Bolivia
  Brazil
  Chile
  Colombia
  Ecuador (including the Galapagos Islands)
  Falkland Islands (Islas Malvinas)
  French Guiana (including Inini)
  Guyana
  Paraguay
  Peru
  Surinam
  Uruguay
  Venezuela

Country Group V

All countries not included in any other country group (except Canada).

Country Group W

Hungary
Poland

Country Group Y

Albania
Bulgaria
Czechoslovakia
Estonia
German Democratic Republic (including East Berlin)
Laos
Latvia
Lithuania
Mongolian People's Republic
Union of Soviet Socialist Republics

Country Group Z

Cambodia
Cuba
North Korea
Vietnam

PART 379

TECHNICAL DATA

§ 379.1

DEFINITIONS[1]

(a) Technical Data[2]

"Technical Data" means information of any kind that can be used, or adapted for use, in the design, production, manufacture, utilization, or reconstruction of articles or materials. The data may take a tangible form, such as a model,[3] prototype,[3] blueprint, or an operating manual (the tangible form may be stored on recording media); or they may take an intangible form such as technical service. All software is technical data.

(b) Export of Technical Data[4,5]

(1) **Export of technical data.** "Export of Technical Data" means—

(i) An actual shipment or transmission of technical data out of the United States[6];

(ii) Any release of technical data in the United States with the knowledge or intent that the data will be shipped or transmitted from the United States to a foreign country; or

(iii) Any release of technical data of U.S.-origin in a foreign country.

(2) **Release of technical data.** Technical data may be released for export through—

(i) Visual inspection by foreign nationals of U.S.-origin equipment and facilities;

(ii) Oral exchanges of information in the United States or abroad; and

(iii) The application to situations abroad of personal knowledge or technical experience acquired in the United States.

(c) Reexport of Technical Data

"Reexport of Technical Data" means an actual shipment or transmission from one foreign country to another, or any release of technical data of U.S. origin in a foreign country with the knowledge or intent that the data will be shipped or transmitted to another foreign country. Technical data may be released for reexport through—

(1) Visual inspection of U.S.-origin equipment and facilities abroad;

(2) Oral exchanges of information abroad; and

(3) The application to situations abroad of personal knowledge or technical experience acquired in the United States.

§ 379.2

LICENSES TO EXPORT

Except as provided in § 370.3(a), an export of technical data must be made under either a U.S. Department of Commerce general license or a validated export license. (See § 371.1 and 372.2 for definitions of "general" and "validated" licenses.) General Licenses *GTDA* and *GTDR* (see § 379.3 and 379.4 below) apply to specific types of exports of technical data. A validated license is required for any export of technical data where these general licenses do not apply, except in the case of certain exports to Canada.[7,8]

§ 379.3

GENERAL LICENSE *GTDA:*
Technical Data Available to all Destinations

A General License designated *GTDA* is hereby established authorizing the export to all destinations of technical data described in § 379.3(a), (b), or (c) below:

(a) Data Generally Available

Data that have been made generally available to the public in any form, including—

---

[1]See § 370.2 for definitions of other terms used in this regulation.
[2]The provisions of Part 379 do not apply to "classified" technical data, i.e. technical data that have been officially assigned a security classification (e.g. "top secret," "secret," or "confidential") by an officer or agency of the U.S. Government. The export of classified technical data is controlled by the Office of Munitions Control of the U.S. Department of State or the U.S. Nuclear Regulatory Commission, Washington, D.C.
[3]Models and prototypes are controlled both as technical data and as commodities. The more restrictive Export Administration requirements apply to their export. See Part 399 for the commodity controls.
[4]License applications for, or questions about, the export of technical data relating to commodities which are licensed by U.S. Government agencies other than the U.S. Department of Commerce shall be referred to such other appropriate U.S. Government agency for consideration (see § 370.10).
[5]Patent attorneys and others are advised to consult the U.S. Patent Office, U.S. Department of Commerce, Washington, D.C. 20231, regarding the U.S. Patent Office regulations concerning the filing of patent applications or amendments in foreign countries. In addition to the regulations issued by the U.S. Patent Office, technical data contained in or related to inventions made in foreign countries or in the United States, are also subject to the U.S. Department of Commerce regulations covering the export of technical data, in the same manner as the export of other types of technical data.
[6]As used in this Part 379, the United States includes its possessions and territories.

[7]Only the restrictions set forth in § 379.4(c) apply to exports of technical data for use in Canada. In all other cases, an export of technical data for use in Canada may be made without either a validated or a general license.
[8]Although Export Administration may provide general information on licensing policies regarding the prospects of approval of various types of export control actions, including actions with respect to technical data, normally it will give a formal judgment respecting a specific request for an action only upon the actual submission of a formal application or request setting forth all of the facts relevant to the export transaction and supported by all required documentation.
Advice is always available, however, regarding any questions as to the applicability of a general license. Such questions should be submitted by letter to the U.S. Department of Commerce, Office of Export Licensing, P.O. Box 273, Washington, D.C. 20044.

(1) Data released orally or visually at open conferences, lectures, trade shows, or other media open to the public; and

(2) Publications that may be purchased without restrictions at a nominal cost, or obtained without costs, or are readily available at libraries open to the public.

The term "nominal cost" as used in § 379.3(a)(2) above, is intended to reflect realistically only the cost of preparing and distributing the publication and not the intrinsic value of the technical data. If the cost is such as to prevent the technical data from being generally available to the public, General License GTDA would not be applicable.

(b) Scientific or Educational Data

(1) Dissemination of information not directly and significantly related to design, production, or utilization in industrial processes, including such dissemination by correspondence, attendance at, or participation in, meetings; or

(2) Instruction in academic institutions and academic laboratories, excluding information that involves research under contract related directly and significantly to design, production, or utilization in industrial processes.

(c) Patent Applications

Data contained in a patent application prepared wholly from foreign-origin technical data where such application is being sent to the foreign inventor to be executed and returned to the United States for subsequent filing in the U.S. Patent and Trademark Office. (No validated export license from the Office of Export Licensing is required for data contained in a patent application, or an amendment, modification, supplement or division thereof for filing in a foreign country in accordance with the regulations of the Patent and Trademark Office in 37 CFR Part 5. See § 370.10(j).)

§ 379.4

GENERAL LICENSE *GTDR:*
Technical Data Under Restriction

A general license designated *GTDR* is hereby established authorizing the export of technical data that are not exportable under the provisions of General License *GTDA*, subject to the provisions, restrictions, exclusions, and exceptions set forth below and subject to the written assurance requirement set forth in § 379.4(f).

(a) Country Groups S and Z Restrictions

No technical data may be exported under this general license to Country Group S or Z.

(b) Restrictions applicable to Country Groups Q, W, Y, T and V, Afghanistan and the People's Republic of China

(1) The following are definitions of terms used to describe certain types of technical data restricted for export to Country Groups Q, W, Y, T and V, Afghanistan and the People's Republic of China;

(i) "Operation technical data" is defined as explicit data in such forms as manuals, instruction sheets, blueprints or software, provided they are:

(A) Sent as part of a transaction involving and directly related to, a commodity licensed for export from the United States, or specifically authorized for reexport, to the same consignee and destination to which the commodity was or will be exported under this license of authorization;

(B) A single shipment sent no later than one year following the shipment of the commodity to which the technical data are related;

(C) Of a type delivered with the commodity in accordance with established business practice;

(D) Necessary to the assembly, installation, maintenance, repair, or operation of the commodity[9]; and

(E) Not related to the production, manufacture, or construction of the commodity.

(ii) "Sales technical data" is defined as data supporting a prospective or actual quotation, bid, or offer to sell, lease, or otherwise supply any commodity, plant or technical data, provided that:

(A) The commodity, plant or technical data are not (and are not related to) a commodity:

(1) Identified by the code letter "A" following the Export Control Commodity Number (ECCN) on the Commodity Control List (Supplement No. 1 to § 399.1), or

(2) Shown in Supplement Nos. 2 or 3 Part 370 (controlled by U.S. Department of State and Nuclear Regulatory Commission, respectively);

(B) The technical data are of a type customarily transmitted with a prospective or actual quotation, bid, or offer (in accordance with established business practice); and

---

[9]Section 376.10(a)(1)(vii) requires exporters of digital computer equipment to describe on their license application any software, including that shipped under General License GTDR, to be used with the equipment.

(C) The export will not disclose the detailed design, production, or manufacture, or the means of reconstruction, of either the quoted item or its product. Similarly, a quotation, bid, or offer for technical data or services must not disclose the detailed technical process involved.

NOTE

Neither this authorization nor its use means that the U.S. Government intends, or is committed, to approve an export license application for any commodity, plant or technical data that may be the subject of the transaction to which such quotation, bid, or offer relates. Exporters are advised to include in any quotations, bids, or offers, and in any contracts entered into pursuant to such quotations, bids, or offers, a provision relieving themselves of liability in the event that an export license (when required) is not approved by the Office of Export Licensing.

(iii) For definitions relating to computer software, see Advisory Note 12 in Supplement No. 3 to Part 379.

(2) No technical data may be exported under this general license to Country Groups Q, W, or Y or to Afghanistan, except:

(i) Software that is not explicitly controlled in Supplement No. 3 to Part 379, or that is not related to any commodity controlled for national security or nuclear non-proliferation reasons on the Commodity Control List, or that is not related to any commodity/technical data listed in paragraphs (c) or (d) of this section.

(ii) "Operation technical data" defined in paragraph (b)(1)(i) of this section, including software in object code explicitly listed in Supp. No. 3 to Part 379 and described on the license application for the commodity, but subject to the restrictions contained in § 379.4(c) and (d).

(iii) "Sales technical data" defined in paragraph (b)(1)(ii) of this section.

(3) The following technical data may be exported under this general license to Country Groups T and V, *except* to the People's Republic of China:

(i) "Sales technical data", as defined in paragraph (b)(1)(ii) of this section, but including "sales technical data" for a commodity identified by the code letter "A" following the Export Control Commodity Number (ECCN) on the Commodity Control List; and

(ii) "Operation technical data" as defined in paragraph (b)(1)(i) of this section, but including "operation technical data", for commodities listed in paragraph (d) of this section.

(4) The following technical data may be exported under this general license to the People's Republic of China:

(i) "Operation technical data" as defined in paragraph (b)(1)(i) of this section, but subject to the restrictions contained in § 379.4(c);

(ii) "Sales technical data" as defined in paragraph (b)(1)(ii) of this section.

(5) Software updates that are intended for and are limited to correction of errors ("fixes" to "bugs" that have been identified) qualify for export under General License *GTDR* without written assurances to Country Groups Q, W, Y, T and V, Afghanistan, and the People's Republic of China, provided that the updates are being exported to the same consignee and do not enhance the functional capabilities of the initial software package.

(c) Technical Data Restrictions Applicable to All Destinations.

No technical data[10] (including operating and maintenance instructional material) related to the following may be exported under this general license, and a validated export license is required for all destinations, *including Canada*, for export of technical data related to the following:

(1) Any commodity where the exporter knows or has reason to know that it will be used directly or indirectly in the following activities, whether or not it is specifically designed or modified for such activities (see § 378.3)—

(i) Designing, developing, fabricating or testing nuclear weapons or nuclear explosive devices,[11,12] or

(ii) Designing, constructing, fabricating, or operating the following facilities, or components for such facilities[13]—

(A) Facilities for the chemical processing of irradiated special nuclear or source material;

(B) Facilities for the production of heavy water;

---

[10]This restriction does not apply to data included in the foreign filing of a patent, provided such foreign filing of a patent application is in accordance with the regulations of the U.S. Patent Office (See § 379.3(c)).

[11]Commodities and technical data specifically designed or specifically modified for use in designing, developing or fabricating nuclear weapons or nuclear explosive devices are subject to export licensing or other requirements of the Office of Munitions Control, U.S. Department of State, or the licensing or other restrictions specified in the Atomic Energy Act of 1954, as amended. Similarly, commodities and technical data specifically designed or specifically modified for use in devising, carrying out, or evaluating nuclear weapons tests or nuclear explosions (except such items as are in normal commercial use for other purposes) are subject to the same requirements.

[12]Also see § 379.5(e) for special provisions relating to technical data for maritime nuclear propulsion plants and other commodities.

[13]Such activities may also require a specific authorization from the Secretary of Energy pursuant to section 57.b.(2) of the Atomic Energy Act of 1954, as amended, as implemented by the Department of Energy's regulations published in 10 CFR 810.

(C) Facilities for the separation of isotopes of source and special nuclear material; or

(D) Facilities for the fabrication of nuclear reactor fuel containing plutonium.

(2) Training of personnel for paragraph (1) of this section.

(3) Neutron generator systems, including tubes, designed for operation without an external vacuum system, and utilizing electrostatic acceleration to induce a tritium-deuterium nuclear reaction; and specially designed parts, n.e.s.

(4) Porous nickel.

(5) Plants specially designed for the production of uranium hexafluoride (UF_6), and specially designed or prepared equipment (including UF_6 purification equipment) and specially designed parts and accessories therefor.

(6) Inverters, converters, frequency changers, and generators having a multiphase electrical power output within the range of 600 to 2000 hertz.

(7) Cylindrical tubing, raw, semifabricated, or finished forms, made of aluminum alloy (7000 series) maraging steel or high-strength titanium alloys (e.g., Ti-6 Al-4 V, etc.) having the following characteristics:

(i) Wall thickness of 1/2 inch, or less; and

(ii) Diameter of 3 inches or more.

(8) Cylindrical rings, or single convolution bellows, made of high-strength steels having all of the following characteristics:

(i) Tensile strength equal to or greater than 150,000 psi;

(ii) Wall thickness of 3 millimeters or less; and

(iii) Diameter of 3 inches or more.

(9) Pipes, valves, fittings, heat exchangers, or magnetic, electrostatic or other collectors made of graphite or stainless steel, or of other materials coated in graphite, yttrium or yttrium compounds resistant to the heat and corrosion of uranium vapor.

(10) Centrifugal balancing machines, fixed or portable, horizontal or vertical, having all of the following characteristics:

(i) Suitable for balancing flexible rotors having a diameter of from 3 inches to 16 inches, and a length of 24 inches or more; and

(ii) Mass capability of from 2 to 50 lbs.; and

(iii) Capable of balancing to a residual imbalance of 0.001 in–lb./lb. per plane or greater; and

(iv) Capable of balancing in three or more planes.

(11) Moisture and particulate separator systems covered by ECCN 1416A(g), as follows:

(i) Technical data for preventing water leakage around the filter stages; and

(ii) Technical data for integrating the components of such a system.

(d) Restrictions Applicable to All Destinations Except Canada

No technical data relating to the following commodities or processes, other than technical data authorized for T and V destinations under paragraphs (b)(3) and (4) of this section,[14] may be exported under this General License GTDR, and exports of these technical data to all destinations, except Canada,[15] require a validated export license—

(1) Civil aircraft, civil aircraft equipment, parts, accessories, or components, except laminated or tempered safety glass for aircraft; hydraulic motors; reciprocating internal combustion engines; air-conditioning systems; heat exchangers and oil and liquid coolers; pumps, air compressors, fans, and blowers; fire extinguishing systems; electric motors and motor controls; electrical apparatus for making, breaking, or protecting electrical circuits; ignition harness and cable sets; electrical starting and ignition equipment; meters and instruments; alarm, warning, and signaling instruments; constant speed propellers, fixed pitch and ground-adjustable propellers for non-military aircraft, and rotors and rotor blades for non-powered rotorcraft, landing lights and other lighting fixtures; apparatus, equipment, and components for oxygen systems; mechanical tachometers, and other aircraft instruments n.e.s. included in the Commodity Control List under No. 6599;

(2) Electrical and electronic instruments specially designed for testing or calibrating the airborne direction finding, navigational, and radar equipment;

(3) Airborne electronic transmitters, receivers, and transceivers;

(4) Airborne electronic direction finding equipment;

---

[14]Data included in the foreign filing of a patent is also excluded from the restrictions set forth in §379.4(d) if such foreign filing of a patent application is in accordance with the regulations of the U.S. Patent Office.

[15]Only the restrictions set forth in §379.4(c) apply to exports of technical data for use in Canada. In all other cases, an export of technical data for use in Canada may be made without either a validated or a general license. For reexport provisions applicable to Canada and other countries, see §379.8(b) and (c).

(5) Airborne electronic or inertial navigation and radar equipment[16];

(6) Watercraft of hydrofoil and hovercraft (air bubble) design;

(7) Submersible watercraft other than military or naval types[17];

(8) Certain airborne multispectral infrared detection and tracking equipment (infrared scanners and their associated technology designed for military or space application remain under the export licensing jurisdiction of the Office of Munitions Control, Department of State);

(9) Infrared imaging equipment;

(10) Technical data specific to the production of "superalloys" (see ECCN 1301A of the Commodity Control List), and technical data specified in operation technical data (§ 379.4(b)(1)(i)) for Country Groups Q, W, Y and Afghanistan, regardless of the export controls on the equipment used with such technical data; melting, remelting and degassing techniques specific to the production of "superalloys"; and technical data specific to the production of "superalloys" in crude and semi-fabricated forms;

(11) Technical data (including processing conditions) and procedures for the regulation of temperature, pressure and/or atmosphere in autoclaves when used for the production of composites or partially processed composites using materials controlled by ECCN 1763A of the Commodity Control List; technical data specified in § 379.4(b)(1)(i) relating to equipment specified in ECCNs 1203A, 4203B, 1312A for Country Groups Q, W, Y and Afghanistan;

(12) Insert gas and vacuum atomizing technology to achieve sphericity and uniform size of particles in metal powders, regardless of the type of metal and the export controls on the powder, and technical data specified in § 379.4(b)(1)(i) relating to equipment specially designed to produce spherical metal powders for Country Groups Q, W, Y and Afghanistan;

(13) Technical data for floating drydocks, limited to the following:

(i) That portion of the design of a floating dock covered by ECCN 1425A(a) that relates to the incorporation of the three types of facilities described in the Note to ECCN 1425A(a); and

(ii) Design, production and use of on-board floating dock facilities covered by ECCN 1425A(b) that permit the operation, maintenance and repair of nuclear reactors;

(14) Military trainer aircraft, specified in ECCN 2460A, except reciprocating aircraft engines;

(15) Military instrument flight trainers;

(16) Technical data for metal-working manufacturing processes and specially designed "software" therefor; (see Supplement No. 4 to Part 379)

(17) "Software" and technical data for "automatically controlled industrial systems" to produce assemblies or discrete parts (see Supplement No. 4 to Part 379);

(18) Databases generated by the use of equipment controlled by ECCN 1363A;

(19) Technical data for application to non-electrical devices to achieve:

(i) Inorganic overlay coatings or inorganic surface modification coatings—(a) specified in column 3 of the Table set forth in Supplement No. 4 to Part 379, (b) on substrates specified in column 2 of that same Table, (c) by processes as defined in Technical Note (a) to (h) and specified in column 1 of that same Table, and specially designed software therefor; and

(20) Design and production technical data for commodities that are controlled for nuclear weapons delivery reasons (see §376.18 and Supplement No. 4 of this part), including related software; and

(21) Any other commodity under the export control jurisdiction of Export Administration, if such commodity is not covered by an entry on the Commodity Control List.

(e) Restrictions Applicable to Republic of South Africa and Namibia

(1) **General Prohibition.** No technical data may be exported or reexported to the Republic of South Africa or Namibia under this General License *GTDR* where the exporter or reexporter knows or has reason to know that the data or any products of the data are for delivery, directly or indirectly, to or for use by or for military or police entities in these destinations or for use in servicing equipment owned, controlled or used by or for such entities. As used in this paragraph (e), the term "any products of the data" includes the direct product[18] of the data and any subsequent products of the direct product. No technical data for use in servicing or manufacturing computers, and no

---

[16]Exports of technical data relating to the design, development, production or manufacture, of inertial navigation equipment, its related parts, components or subsystems, and exports of technical data relating to the repair of parts, components or subsystems of inertial navigational systems (including accelerometers and gyroscopes) not certified by the FAA or not used as an integral part of civil aircraft, are controlled by the Office of Munitions Control, Department of State.

[17]Technical data relating to military or naval submersible watercraft are subject to the export licensing authority of the U.S. Department of State.

[18]The term "direct product," as used in this paragraph, is defined to mean the immediate product (including processes and services) produced directly by use of the technical data.

computer software, may be exported or reexported to the Republic of South Africa or Namibia under this General License GTDR where the exporter or reexporter knows or has reason to know that the data will be made available to or for use by, or is intended to be used for, apartheid-enforcing entities identified in Supplement No. 1 to Part 385. In addition, no technical data relating to the commodities listed in Supplement No. 2 to this Part 379 may be exported or reexported under this General License *GTDR* to any consignee in the Republic of South Africa or Namibia.

(2) **Written Assurance.** In addition to any written assurances that may be required by paragraph (f) of this section, no export or reexport of technical data, including computer software, may be made to the Republic of South Africa or Namibia under this General License GTDR until the exporter has received written assurance from the importer that neither the technical data nor the direct product of the data will be made available to or for use by or for military or police entities of the Republic of South Africa or Namibia. If the technical data is intended to service or manufacture computers or consists of computer software, the written assurance must also state that the data will not be made available to or for use by, and neither the data nor the direct product of the data is intended to be used for, the apartheid-enforcing entities identified in Supplement No. 1 to Part 385. To facilitate this assurance, the potential exporter shall provide a current copy of Supplement No. 1 to Part 385 to the intended recipient at the time the written assurance is requested.

(f) Written Assurance Requirements

(1) **Requirement of written assurance for certain data, services, and materials.** No export of technical data of the kind described in paragraphs (f)(1)(i)(A) through (Q) (not (R)) of this section may be made under the provisions of this General License *GTDR* until the exporter has received written assurance from the importer that neither the technical data nor the direct product[19] thereof is intend-

ed to be shipped, either directly or indirectly, to Country Group Q, S, W,[20] Y, or Z, or Afghanistan or the People's Republic of China, except as provided in paragraph (f)(1)(ii) of this section. No export of technical data of the kind described in paragraph (f)(1)(i)(R) may be made under the provisions of this General License *GTDR* until the exporter has received written assurance from the importer that neither the technical data nor the direct product[19] thereof is intended to be shipped, directly or indirectly, to the Kama River (Kam AZ) or ZIL truck plants in the U.S.S.R., except as provided in paragraph (f)(1)(ii) of this section. The required assurance may be in the form of a letter or other written communication from the importer evidencing such intention, or a licensing agreement that restricts disclosure of the technical data to use only in a country other than Country Group Q, S, W, Y, or Z, or Afghanistan or the People's Republic of China, and prohibits shipments of the direct product[19] thereof by the licensee to Country Group Q, S, W, Y, or Z, or Afghanistan or the People's Republic of China or, for data of the kind described in paragraph 379.4(f)(1)(i)(R), to the Kama River (Kam AZ) or ZIL truck plants in the U.S.S.R. An assurance included in a licensing agreement will be acceptable for all exports made during the life of the agreement. If such assurance is not received, this general license is not applicable and a validated export license is required. An application for such validated license shall include an explanatory statement setting forth the reasons why such assurance cannot be obtained. In addition, this general license is not applicable to any export of technical data of the kind described in paragraphs (f)(1)(i)(A) through (Q) (not (R)) of this section if, at the time of export of the technical data from the United States, the exporter knows or has reason to believe that the direct product to be manufactured abroad by use of the technical data is intended to be exported or reexported, directly or indirectly, to Country Group Q, S, W, Y, or Z, or Afghanistan or the People's Republic of China, or, for data of the kind described in paragraph 379.4(f)(1)(i)(R), to the Kama River (Kam AZ) or ZIL truck plants in the U.S.S.R.

(i) Technical data relating to the following commodities—

---

[19] The term "direct product," as used in this sentence and in this context only, is defined to mean the immediate product (including processes and services) produced directly by use of the technical data, except that petroleum or chemical products other than molecular sieves or catalysts are not included in this definition. The coverage of the term does not extend to the results of the use of such "direct product." An example of the direct product of technical data is reforming process equipment designed and constructed by use of the technical data exported, but the aromatics produced by the reforming process equipment are not immediate or direct products of these technical data. However, if the technical data are a formula for producing aromatics, the aromatics, although they are immediate products of the data, are not included in this definition of direct product, since they are petroleum products. Conversely, if the technical data are a formula for producing either molecular sieves or catalysts, the foreign-produced molecular sieves and catalysts are included in the definition of direct product.

[20] Effective April 26, 1971, Country Group W no longer included Romania. Assurances executed prior to April 26, 1971, and referring to Country Group W continue to apply to Romania as well as Poland. Effective June 2, 1980, Hungary was added to Country Group W, which at that time included only Poland. Assurances executed prior to June 2, 1980 and referring to Country Group Y continue to apply to Hungary. Assurances executed on or after June 2, 1980 and referring to Country Group W apply to Hungary as well as Poland.

(A) Electronic computers, if the technical data is software listed in Supplement No. 3 to Part 379. (Software not listed in Supplement No. 3 to Part 379 or any ECCN entry, or in § 379.4(c) or (d) may be exported to Country Groups T & V, without written assurance.)

(B) Pyrolitically-derived materials formed on a mold, mandrel or other substrate from precursor gases that decompose in the 1,573K (1,300° C) to 3.173K (2,900° C) temperature range at pressures of 133.3 Pa to 19.995 kPa (including the composition of precursor gases, flow rates, and process control schedules and parameters);

(C) Other gravity meters (gravimeters); and parts and accessories, n.e.s.;

(D) Other transonic, supersonic, hypersonic and hypervelocity wind tunnels and devices; and parts and accessories, n.e.s.;

(E) Watercraft of 65 feet and over in overall length, designed to include motors or engines of 600 horsepower or over and greater than 45 displacement tons;

(F) Methyl methacrylate, cross-linked, hot stretched, clear, film, sheeting, or laminates;

(G) Doppler sonar navigation systems;

(H) Aerial camera film, sensitized and unexposed, as follows—

(1) Having spectral sensitivities at wavelengths greater than 7,500 Angstroms or at wavelengths less than 2,000 Angstroms; or

(2) Having resolving powers (using a Test-Object Contrast of 1,000:1) of 200 line pairs/mm or more or with a base thickness before coating of 0.0025 inch or less;

(I) Continuous tone aerial duplicating film, sensitized and unexposed, having resolving powers (using a Test-Object Contrast of 1,000:1) of 300 line pairs/mm or more;

(J) Instrumentation and/or recording film, sensitized and unexposed, having photo-recording sensitivities (as based on the reciprocal of the tungsten exposure in meter-candle-seconds at an exposure time of 0.0001 second) of 125 or more and a resolving power (using a Test-Object Contrast of 1,000:1) of 55 line pairs/mm or more and with a base thickness before coating of 0.004 inch or less and capable of being processed in solutions with alkalinities of pH 10 or above at temperatures greater than 85° F;

(K) Technical data/software specifically identified in ECCN 1391A. (Operation software defined by

§ 379.4(b)(1) shall be described in the license applications for these commodities.);

(L) Technical data for the design and production (except assembly and testing) of two-axis numerical control units with an "embedded" computer;

(M) Other high speed continuous writing, rotating drum cameras capable of recording at rates in excess of 2,000 frames per second; and parts and accessories, n.e.s.;

(N) Other 16 mm high-speed motion picture cameras capable of recording at rates in excess of 2,000 frames per second; and parts and accessories, n.e.s.;

(O) Single crystal sapphire substrates:

(P) [Reserved]

(Q) Terminal, multiplex or modem equipment designed for or used as components, accessories or sub-assemblies of frequency, time or space division telephone-switching systems employing digital transmission techniques designed at a data signaling rate of 2.1 megabits or less.

(R) Commodities used directly in production, assembly, or testing of trucks, truck components, or accessories at the Kama River (Kam AZ) or ZIL truck plants in the U.S.S.R.

(ii) The limitations set forth in this § 379.4 (f)(1) do not apply to the export of—

(A) Technical data included in an application for the foreign filing of a patent, provided such foreign filing of a patent application is in accordance with the regulations of the U.S. Patent Office; and

(B) Sales technical data supporting a price quotation as authorized in § 379.4 (b)(3) and (4) above.

(2) Requirement of written assurance for certain additional products and destinations.

(i) Except for technical data requiring a written assurance in accordance with the provisions of § 379.4 (f)(1) above, and except as provided in § 379.4 (f)(2)(v) below, no export of technical data relating to the commodities described below in this § 379.4 (f)(2) may be made under the provisions of this General License GTDR until the U.S. exporter has received a written assurance from the foreign importer that, unless prior authorization is obtained from the Office of Export Licensing, the importer will not knowingly—

(A) Reexport, directly or indirectly, to Country Group Q, S, W,[21] Y, or Z, or Afghanistan or the People's Republic of China any technical data relating to commodities identified by the symbol "W" in the paragraph of any entry on the Commodity Control List titled "Validated License Required";

(B) Export, directly or indirectly, to Country Group Z any direct product[22] of the technical data if such direct product is identified by the symbol "W" in the paragraph of any entry on the Commodity Control List titled "Validated License Required;" or

(C) Export, directly or indirectly, to any destination in Country Group Q, S, W,[21] Y or Afghanistan or the People's Republic of China, any direct product of the technical data if such direct product is identified by the code letter "A" following the Export Control Commodity Number on the Commodity Control List.

(ii) If the direct product[23] of any technical data is a complete plant or any major component of a plant that is capable of producing a commodity identified by the symbol "W" in the paragraph of any entry on the Commodity Control List titled "Validated License Required" or appears in the U.S. Munitions List, a written assurance by the person who is or will be in control of the distribution of the products of the plant (whether or not such person is the importer) shall be obtained by the U.S. exporter (via the foreign importer), stating that, unless prior authorization is obtained from the Office of Export Administration, such person will not knowingly—

(A) Reexport, directly or indirectly, to Country Group Q, S, W,[21] Y, Z, or Afghanistan or the People's Republic of China, the technical data relating to the plant or the major component of a plant;

(B) Export, directly or indirectly, to Country Group Z the plant or the major component of a plant (depending upon which is the direct product[22,23] of the technical data) or any product of such plant or of such major component if such product of the plant is identified by the symbol "W" in the paragraph of any entry on the Commodity Control List titled "Validated License Required" or appears in the U.S. Munitions List; or

(C) Export, directly or indirectly, to Country Group Q, S, W, Y, or Afghanistan or the People's Republic of China, the plant or the major component of a plant (depending upon which is the direct product of the technical data), or any product of such plant or of such major component, if such product is identified by the code letter "A" following the Export Control Commodity Number on the Commodity Control List, or appears in the U.S. Munitions List.

NOTE

Effective April 1, 1964 § 379.4(f)(2)(ii) *(B)* and *(C)* required certain written assurances relating to the disposition of the products of a complete plant or major component of a plant which is the direct product of unpublished technical data of U.S. origin exported under General License *GTDR.*

Except as to commodities identified by the code letter "A" following the Export Control Commodity Number on the Commodity Control List, and items on the U.S. Munitions List, the effective date of the written assurance requirements for plant products as a condition of using General License *GTDR* for export of this type of technical data is hereby deferred until further notice, subject to the following limitations:

1. The exporter shall, at least two weeks before the initial export of the technical data, notify the Office of Export Licensing, by letter, of the facts required to be disclosed in an application for a validated export license covering such technical data; and

2. The exporter shall obtain from the person who is or will be in control of the distribution of the products of the plant (whether or not such person is the importer) a written commitment that he will notify the U.S. Government, directly or through the exporter, whenever he enters into negotiations to export any product of the plant to any destination covered by § 379.4(f)(2)(ii)(B) of this part, when such product is not identified by the code letter "A" following the Export Control Commodity Number on the Commodity Control List and requires a validated license for export to Country Group W by the information set forth in the applicable CCL entry in the paragraph titled "Validated License Required." The notification should state the product, quantity, country of destination, and the estimated date of shipment.

Moreover, during the period of deferment, the remaining written assurance requirements of § 379.4(f)(2)(ii) *(B)* and *(C)* as to plant products that are identified by the code letter "A" following the Export Control Commodity Number on the Commodity Control List, or are on the U.S. Munitions List, will be waived if the plant is located in one of the following COCOM countries: Belgium, Canada, Denmark, the Federal Republic of Germany, France, Greece, Italy, Japan, Luxembourg, The Netherlands, Norway, Portugal, Spain, Turkey, and the United Kingdom. This deferment applies to exports of technical data pursuant to any type of contract or arrangement, including licensing agreements, regardless of whether entered into before or after April 1, 1964.

(iii) The required assurance may be in the form of a letter or other written communication from the

---

[21] Effective April 26, 1971, Country Group W no longer included Romania. Assurances executed prior to April 26, 1971, and referring to Country Group W continue to apply to Romania as well as Poland. Effective June 2, 1980, Hungary was added to Country Group W, which at that time included only Poland. Assurances executed prior to June 2, 1980, and referring to Country Group Y continue to apply to Hungary. Assurances executed on or after June 2, 1980 and referring to Country Group W apply to Hungary as well as Poland.

[22,23] The term "direct product" as used in this sentence and in this context only, is defined to mean the immediate product (including processes and services) produced directly by use of the technical data.

importer or, if applicable, the person in control of the distribution of the products of a plant; or the assurance may be incorporated into a licensing agreement which restricts disclosure of the technical data to use only in authorized destinations, and prohibits shipment of the direct product[24] thereof by the licensee to any unauthorized destination. An assurance included in a licensing agreement will be acceptable for all exports made during the life of the agreement. If such assurance is not received, this general license is not applicable and a validated export license is required. An application for such validated license shall include an explanatory statement setting forth the reasons why such assurance cannot be obtained.

(iv) In addition, this general license is not applicable to any export of technical data of the kind described in this § 379.4(f)(2) if, at the time of export of the technical data from the United States, the exporter knows or has reason to believe that the direct product[24] to be manufactured abroad by use of the technical data is intended to be exported directly or indirectly to any unauthorized destination.

(v) The limitations set forth in this § 379.4(f)(2) do not apply to the export of—

(A) Technical data included in an application for the foreign filing of a patent provided such foreign filing of a patent application is in accordance with the regulations of the U.S. Patent Office;

(B) Technical data supporting a price quotation as described in § 379.4 (b)(2) above; and

(C) Technical data relating to those commodities listed in a Supplement to Part 377 as being under short supply control.

(3) Requirement of written assurances for entry No. 1572 on the Commodity Control List (CCL). No technical data related to CCL entry 1572A, Exceptions 2 through 4, may be exported under the provisions of this General License GTDR until the exporter has received written assurance from the importer that the technical data will not be shipped, either directly or indirectly, to Country Groups Q, S, W, Y or Z, or Afghanistan or the People's Republic of China. The letter of assurance requirements are stated in § 379.4(f)(1).

---

[24]The term "direct product" as used in this sentence and in this context only, is defined to mean the immediate product (including processes and services) produced directly by use of the technical data.

NOTE

A written assurance is not required for the export under this General License GTDR of any technical data which do not fall within the description set forth in § 379.4(f)(1) or (2) above.

(g) Restrictions Applicable to Software

Software and other technical data listed under ECCN 1391A on the Commodity Control List may be exported or reexported under General License GTDR only to Country Groups T & V, except Afghanistan and the People's Republic of China, subject to the written assurance requirements contained in § 379.4(f)(1)(i)(K). Such software and technical data require a validated license for export to any other Country Group. The export or reexport of all other software and technical data listed under any other ECCN on the Commodity Control List requires a validated license to any Country Group.

(h) Restrictions Applicable to Technical Data Related to Crime Control and Detection Commodities

(1) The export of technical data related to crime control and detection instruments and equipment under General License GTDR is authorized only to Australia, Belgium, Canada, Denmark, France, the Federal Republic of Germany (including West Berlin), Greece, Iceland, Italy, Japan, Luxembourg, the Netherlands, New Zealand, Norway, Portugal, Spain, Turkey, and the United Kingdom. An individual validated export license is required for the export of such technical data to any other destination, unless the export is authorized under General License GTDA (§ 379.3). (See § 376.14 for a list of the Commodity Control List items affected by the validated license requirement).

(i) Additional restrictions applicable to the People's Republic of China

In addition to the prohibitions in § 379.4 (c) and (d), no technical data related to commodities identified on the Commodity Control List (Supplement No. 1 to § 399.1) as controlled for reasons of national security, nuclear non-proliferation, or crime control may be exported to the People's Republic of China under General License GTDR. In addition, software covered by Supplement 3 to Part 379 requires a validated license for export to the PRC. These prohibitions do not apply, however, to technical data described in § 379.4(b).

§ 379.5

VALIDATED LICENSE APPLICATIONS

(a) General

No technical data, other than that exportable without license to Canada or under general license to other destinations, may be exported from the United States without a validated export license. Such

(b) Application Form

Form ITA–622P shall be completed as provided in § 372.4, except that Items 9(a), 9(c) and 11 shall be left blank. In Item 9(b), "Description of Commodity or Technical Data," enter a general statement which specifies the technical data (e.g., blueprints, manuals, etc.). In addition, the words "Technical Data" shall be entered in Item 4, "Special Purpose."

(c) [Reserved.]

(d) Letter of Explanation.

Each application shall be supported by a comprehensive letter of explanation in duplicate. This letter shall set forth all the facts required to present to the Office of Export Licensing a complete disclosure of the transaction including, if applicable, the following—

(1) The identities of all parties to the transaction;

(2) The exact project location where the technical data will be used;

(3) The type of technical data to be exported;

(4) The form in which the export will be made;

(5) The uses for which the data will be employed;

(6) An explanation of the process, product, size, and output capacity of the equipment, if applicable, or other description that delineates, defines, and limits the data to be transmitted (the "technical scope");

(7) The availability abroad of comparable foreign technical data.

(e) Special Provisions

(1) **Maritime nuclear propulsion plants and related commodities.**[25] These special provisions are applicable to technical data relating to maritime (civil) nuclear propulsion plants, their land prototypes, and special facilities for their construction, support, or maintenance, including any machinery, device, component, or equipment specifically developed or designed for use in such plants or facilities. Every application for license to

---

[25] See § 379.8(a) which sets forth provisions prohibiting exports and reexports of certain technical data and products manufactured therefrom.

validated export licenses are issued by the Office of Export Licensing upon receipt of an appropriate export application or reexport request. An application for a technical data license shall consist of—

(1) Form ITA–622P, Application for Export License, accompanied by

(2) A letter of explanation described in § 379.5 (d). export technical data relating to any of these commodities shall include the following—

(i) A description of the foreign project for which the technical data will be furnished;

(ii) A description of the scope of the proposed services to be offered by the applicant, his consultant(s), and his subcontractor(s), including all the design data which will be disclosed;

(iii) The names, addresses and titles of all personnel of the applicant, his consultant(s) and his subcontractor(s) who will discuss or disclose the technical data or be involved in the design or development of the technical data;

(iv) The beginning and termination dates of the period of time during which the technical data will be discussed or disclosed and a proposed time schedule of the reports which the applicant will submit to the U.S. Department of Commerce, detailing the technical data discussed or disclosed during the period of the license;

(v) The following certification:

I (We) certify that if this application is approved, I (we) and any consultants, subcontractors, or other persons employed or retained by us in connection with the project thereby licensed will not discuss with or disclose to others, directly or indirectly, any technical data relating to U.S. naval nuclear propulsion plants. I (We) further certify that I (we) will furnish to the U.S. Department of Commerce all reports and information which it may require concerning specific transmittals or disclosures of technical data pursuant to any license granted as a result of this application.

(vi) A statement of the steps which the applicant will take to assure that personnel of the applicant, his consultant(s) and his subcontractor(s) will not discuss or disclose to others technical data relating to U.S. naval nuclear propulsion plants; and

(vii) A written statement of assurance from the foreign importer that unless prior authorization is obtained from the Office of Export Licensing, the importer will not knowingly export directly or indirectly to Country Group Q, W,[26] Y, or Z, or the

---

[26] Effective April 26, 1971, Country Group W no longer included Romania. Assurances executed prior to April 26, 1971, and referring to Country Group W continue to apply to Romania as well as Poland. Effective June 2, 1980, Hungary was added to Country Group W, which at that time included only Poland. Assurances executed prior to June 2, 1980 and referring to Country Group Y continue to apply to

People's Republic of China or Afghanistan the direct product of the technical data. However, if the U.S. exporter is not able to obtain this statement from the foreign importer, the U.S. exporter shall attach an explanatory statement to his license application setting forth the reasons why such an assurance cannot be obtained.

(2) **Other commodities.** For all license applications to export to any destination, other than Country Group Q, W, Y, or Z, the People's Republic of China or Afghanistan, technical data relating to any of the commodities set forth below, an applicant shall attach to the license application a written statement from his foreign importer assuring that, unless prior authorization is obtained from the Office of Export Licensing, the importer will not knowingly reexport the technical data to any destination, or export the direct product of the technical data, directl y or indirectly, to Country Group Q, W, Y, or Z, the People's Republic of China or Afghanistan. However, if the U.S. exporter is not able to obtain the required statement from his importer, the exporter shall attach an explanatory statement to his license application setting forth the reasons why such an assurance cannot be obtained. The special provisions set forth in this § 379.5(e)(2) are applicable to technical data concerning the following—

(i) Commodities where the exporter knows or has reason to know that the item will be used directly or indirectly, whether or not specifically designed, for developing or testing nuclear weapons or nuclear explosive devices, nuclear testing, the chemical processing of irradiated special nuclear or source material, the production of heavy water, the separation of isotopes of source and special nuclear material, or the fabrication of nuclear reactor fuel containing plutonium, as described in § 378.3, or training of personnel for any activity listed above;

(ii) Neutron generator systems, including tubes, designed for operation without an external vacuum system, and utilizing electrostatic acceleration to induce a tritium-deuterium nuclear reaction; and specially designed parts, n.e.s.;

(iii) Porous nickel;

(iv) Civil aircraft, civil aircraft equipment, parts, accessories, or components identified by the code letter "A," "B," or "M" following the Export Control Commodity Number on the Commodity Control List (Supplement No. 1 to § 399.1);

(v) Electrical and electronic instruments specially designed for testing or calibrating the airborne direction finding, navigational, and radar equipment;

(vi) Airborne electronic transmitters, receivers, and transceivers;

(vii) Airborne electronic direction finding equipment;

(viii) Airborne electronic navigation and radar equipment;

(ix) Watercraft of hydrofoil and hovercraft (air bubble) design[27];

(x) Submersible watercraft other than military or naval types[28];

(xi) Plants specially designed for the production of uranium hexafluoride (UF_6), and specially designed or prepared equipment (including UF_6 purification equipment) and specially designed parts and accessories therefor;

(xii) Inverters, converters, frequency changers, and generators having a multiphase electrical power output within the range of 600 to 2000 hertz;

(xiii) Cylindrical tubing, raw, semifabricated, or finished forms, made of aluminum alloy (7000 series) maraging steel or high-strength titanium alloys (e.g., Ti–6 Al–4 V, etc.) having the following characteristics:

(A) Wall thickness of 1/2 inch, or less;

(B) Diameter of 3 inches or more;

(xiv) Cylindrical rings, or single convolution bellows, made of high-strength steels having all of the following characteristics:

(A) Tensile strength equal to or greater than 150,000 psi;

(B) Wall thickness of 3 millimeters or less; and

(C) Diameter of 3 inches or more;

(xv) Pipes, valves, fittings, heat exchangers, or magnetic, electrostatic or other collectors made of graphite or stainless steel, or of other materials coated in graphite, yttrium or yttrium compounds resistant to the heat and corrosion of uranium vapor; and

(xvi) Any other commodity under the export control jurisdiction of Export Administration if

---

[27] This commodity is not listed on the Commodity Control List since it is under the export control jurisdiction of the U.S. Maritime Administration. However, technical data relating to this commodity are under the export control jurisdiction of Export Administration.

[28] Technical data relating to military or naval submersible watercraft are subject to the export licensing authority of the U.S. Department of State. See Supplement No. 2 to Part 370.

Hungary. Assurances executed on or after June 2, 1980 and referring to Country Group W apply to Hungary as well as Poland.

such commodity is not covered by an entry on the Commodity Control List.

(3) Computer software. Applications to export software subject to validated export licenses should provide the information detailed in § 376.10 (a)(1)(vii) and, in addition, should identify the location, origin, and model of the equipment with which the software will be used. For software, the applicant should indicate on the application the Commodity Control List Processing Code for the equipment with which the software will be used.

(f) Validity Period and Extension

(1) Initial validity. Validated licenses covering exports of technical data will generally be issued for a validity period of 24 months. Upon request, a validity period exceeding 24 months may be granted where the facts of the transaction warrant it and the Office of Export Licensing determines that such action would be consistent with the objectives of the applicable U.S. export control program. Justification for a validity period exceeding 24 months should be provided in accordance with the procedures set forth in § 372.9(d)(2) for requesting an extended validity period with a license application. The Office of Export Licensing will make the final decision on what validity beyond 24 months, if any, should be authorized in each case.

(2) Extensions. A request to extend the validity period of a technical data license shall be made on Form ITA-685P in accordance with the procedures set forth in § 372.12(a). The request shall include on Form ITA-685P, in the space entitled "Amend License to Read as Follows," whether the license has been previously extended and the date(s) and duration of such extension(s). The Office of Export Licensing will make the final decision on what extension beyond 24 months, if any, should be authorized in each case. (See § 379.8(c)(1) for validity period extensions for reexports of technical data.)

§ 379.6

EXPORTS UNDER A VALIDATED LICENSE

(a) Use of Validated Licenses

(1) Retention of license. The validated technical data license need not be presented to the customs office or post office but shall be retained and made available for inspection in accordance with the recordkeeping provisions of § 387.13.

(2) Return of license. Export licenses shall be returned promptly to the Office of Export Licensing upon revocation, suspension, or expiration of the validity period. Used licenses shall be returned when fully used. Unused and partially used licenses shall be returned when the exporter determines that he will not make any shipment, or any further shipment, thereunder, or upon expiration, whichever comes first.

(b) Reports on Exports

(1) Country Group S, T, or V except Afghanistan. With respect to a license used to export technical data to Country Group S, T, or V except Afghanistan when the license is returned, as provided in § 379.6(a)(2) above, the exporter shall submit a statement indicating—

(i) When the technical data were exported or when the technical services were rendered; and

(ii) Whether the export was total or partial.

(iii) For exports to the People's Republic of China, the exporter also shall submit a statement indicating the nature of the transaction (*e.g.*, a sale of technical data, performance of technical services, a technical licensing agreement, or a technology exchange agreement).

(2) Country Group Q, W, or Y or Afghanistan. With respect to a license used to export technical data to Country Group Q, W, or Y or Afghanistan, when the license is returned, as provided in § 379.6(a)(2) above, the exporter shall submit a statement indicating—

(i) When the technical data were exported or when the technical services were rendered;

(ii) Whether the export or service was total or partial;

(iii) The nature of the transaction (*e.g.*, a sale of technical data, performance of technical services, a technical licensing agreement, a technology exchange agreement);

(iv) The nature of the payment received or to be received by the U.S. exporter (*e.g.*, pecuniary or other consideration); and

(v) The actual or estimated price of the technical data exported, or services rendered, or the actual or estimated dollar value of any other consideration received or to be received. (This should include the payment received or to be received for engineering and for any other services when rendered, as well as for the royalty or other payment received or to be received for a design or process authorized to be used.)

§ 379.7

AMENDMENTS

Requests for amendments shall be made in accordance with the provisions of § 372.11. Changes requiring amendment include any expansion or upgrade of the technical scope that was described in the letter of explanation, as approved or modified on the export license.

§ 379.8

REEXPORTS OF TECHNICAL DATA AND EXPORTS OF THE PRODUCT MANUFACTURED ABROAD BY USE OF U.S. TECHNICAL DATA

(a) Prohibited Exports and Reexports

Unless specifically authorized by the Office of Export Licensing, or otherwise authorized under the provisions of § 379.8(b) below, no person in the United States or in a foreign country may—

(1) Reexport any technical data imported from the United States, directly or indirectly, in whole or in part, from the authorized country(ies) of ultimate destination;

(2) Export any technical data from the United States with the knowledge that it is to be reexported, directly or indirectly, in whole or in part, from the authorized country(ies) of ultimate destination; or

(3) Export or reexport to Country Group Q, W, Y, or Z, the People's Republic of China or Afghanistan any foreign produced direct product of U.S. technical data, or any commodity produced by any plant or major component thereof which is a direct product of U.S. technical data, if such direct product or commodity is covered by the provisions of § 379.4(f) or § 379.5(e) (1) or (2).

(b) Permissive Reexports

(1) **Exportable under General License *GTDA* or *GTDR*.** Any technical data which have been exported from the United States may be reexported from any destination to any other destination, provided that, at the time of reexport, the technical data may be exported directly from the United States to the new country of destination under General License *GTDA* or *GTDR*, and provided that all of the requirements and conditions for use of these general licenses have been met.

(2) **Country Groups Q, W, and Y and Afghanistan.** When the Office of Export Licensing has specifically authorized the export of a commodity from the United States to a destination in Country Group Q, W, or Y or Afghanistan or the reexport of a U.S.-origin commodity from any foreign country to a destination in Country Group Q, W, or Y or Afghanistan, technical data such as manuals, instructional sheets, or blueprints as described in, and subject to the conditions of, § 379.4(b) (1) may be sent to the same destination as part of the same transaction without separate specific authorization by the Office of Export Licensing.

(3) **COCOM Authorization.** Separate specific authorization by the Office of Export Licensing to export or reexport any foreign-produced direct product of U.S. technical data is not required if *all* of the following conditions are met:

(i) The commodities being exported:

(A) Are identified by an "A" on the CCL,

(B) Are *not* included in the Advisory Notes in any entry on the Commodity Control List (Supplement No. 1 to § 399.1) (the Advisory Notes for the People's Republic of China do *not* apply in meeting this requirement),

(C) Are valued at more than $4,000 (except that commodities in CCL entries 1548, 1555, 1559, and 1781, and entries beginning with the digits 2 and 3, are eligible regardless of value), and

(D) If included in entry 1565A, have at least one parameter falling in the extreme right-hand, *i.e.* (d), column of boxes on Form ITA–6031P, Computer System Parameters.

(ii) The export or reexport is from a COCOM participating country, *i.e.*, Belgium, Canada, Denmark, France, the Federal Republic of Germany, Greece, Italy, Japan, Luxembourg, The Netherlands, Norway, Portugal, Spain, Turkey, or the United Kingdom;

(iii) The export or reexport is made in accordance with the conditions of the licensing authorization issued by the applicable COCOM participating country;

(iv) The export or reexport is to a country in Country Group Q, W, or Y or the People's Republic of China; and

(v) The transaction has received unanimous approval in COCOM (see § 374.3(e)).

(4) **People's Republic of China.** Separate specific authorization by Export Administration is not required to reexport software from a COCOM country to the People's Republic of china if the software meets the requirements set forth in the Advisory Note for the People's Republic of China in Supple-

ment No. 3 to Part 379 and if the license has been approved by a COCOM country.

(c) Specific Authorization To Reexport

(1) Submission of request for reexport authorization. Requests for specific authorization to reexport technical data or to export any product thereof, as applicable, shall be submitted on Form ITA–699P, Request To Dispose of Commodities or Technical Data Previously Exported, to:

> Office of Export Licensing
> P.O. Box 273
> Washington, D.C. 20044.

(See Supplement No. 1 to Part 374 for instructions on completing the form.) If Form ITA–699P is not readily available, a request for specific authorization to reexport technical data or to export any product thereof, as applicable, may be submitted by letter. The letter shall bear the words "Technical Data Reexport Request" immediately below the heading or letterhead and contain all the information required by § 379.5(d). Authorization to reexport technical data or to export the product thereof, if granted, will generally be issued with a validity period of 24 months on Form ITA–699P, or by means of a letter from the Office of Export Licensing.

Any request for extension of the validity period shall be requested in accordance with § 374.5(b), and shall specify the period for which additional validity is required. The Office of Export Licensing will make the final decision on what validity beyond 24 months, if any, should be authorized in each case.

(2) Return of reexport authorization. Reexport authorizations shall be returned promptly to the Office of Export Licensing upon revocation, suspension, or expiration of the validity period. Used authorizations shall be returned when fully used. Unused and partially used authorizations shall be returned when the person authorized to reexport determines that he will not make any shipment, or further shipment, thereunder, or upon expiration of the authorization, whichever comes first. After the reexport of the technical data has been completed, the Office of Export Licensing shall also be given a notice in writing indicating—

(i) When the technical data were reexported or when the technical services were rendered; and

(ii) Whether the reexport or service was total or partial.

(iii) For exports to the People's Republic of China, a notice also is required indicating the nature of the transaction (*e.g.*, a sale of technical data, performance of technical services, a technical licensing agreement, a technology exchange agreement, or the rendering of technical services).

(3) Reexports to Country Group Q, W, or Y or Afghanistan. In addition, if the technical data had been reexported to Country Group Q, W, or Y or Afghanistan, the written notice shall specify—

(i) The nature of the transaction *e.g.*, a sale of technical data, performance of technical services, a technical licensing agreement, a technology exchange agreement, or the rendering of technical services);

(ii) The nature of the payment received, or to be received, by the U.S. exporter (*e.g.*, pecuniary or other consideration); and

(iii) The actual or estimated price of the technical data reexported or services rendered, or the actual or estimated dollar value of any other consideration received or to be received. (This should include the payment received or to be received for engineering and for any other services when rendered, as well as for the royalty or other payment received or to be received for a design or process authorized to be used.)

(d) Effect of Foreign Laws

No authority granted by the U.S. Office of Export Licensing, or under the provisions of the U.S. Export Administration Regulations, to reexport technical data or export a product thereof shall in any way relieve any person from his responsibility to comply fully with the laws, rules, and regulations of the country from which the reexport or export is to be made or of any other country having authority over any phase of the transaction. Conversely, no foreign law, rule, regulation, or authorization in any way relieves any person from his responsibility to obtain such authorization from the U.S. Office of Export Licensing as may be required by the U.S. Export Administration Regulations.

§ 379.9

COMMERCIAL AGREEMENTS WITH CERTAIN COUNTRIES

Pursuant to section 5(j) of the Export Administration Amendments Act of 1979, as amended, any non-governmental U.S. person or firm that enters into an agreement with any agency of the government of a controlled country (Country Groups Q, W, Y, and the People's Republic of China), which agreement encourages technical cooperation and is intended to result in the export from the U.S. to the other party of U.S.-origin technical data (except under General License GTDA or General license GTDR as provided under the provisions of § 379.4(b)(1) and (2) of this part), shall submit those portions of the agreement that include the statement of work and describe the anticipated exports of data to the Office of Technology and Policy Analysis, Room 4054, P.O. Box 273, Washington, D.C. 20044. This material shall be submitted no later than 30 days after the final signature on the agreement.

(a) This requirement does not apply to colleges, universities and other educational institutions.

(b) The submission required by this section does not relieve the exporter from the licensing requirements for controlled technical data and goods.

(c) Acceptance of a submission does not represent a judgment as to whether Export Administration will or will not issue any authorization for export of technical data.

§ 379.10

OTHER APPLICABLE PROVISIONS

As far as may be consistent with the provisions of this Part 379, all of the other provisions of the Export Administration Regulations shall apply equally to exports of technical data and to applications for licenses and licenses issued under this part.

TECHNICAL DATA INTERPRETATIONS

1. Technology Based on U.S.-Origin Technical Data

U.S.-origin technical data does not lose its U.S.-origin when it is redrawn, used, consulted, or otherwise commingled abroad in any respect with other technical data of any other origin. Therefore, any subsequent or similar technical data prepared or engineered abroad for the design, construction, operation, or maintenance of any plant or equipment, or part thereof, which is based on or utilizes any U.S.-origin technical data, is subject to the same U.S. Export Administration Regulations that are applicable to the original U.S.-origin technical data, including the requirement for obtaining Office of Export Licensing authorization prior to reexportation.

2. Distinction Between General and Validated License Requirements for Shipment to QWY Destinations of Technical Data and Replacement Parts

A number of exporters have recently asked where the line is drawn between general license and validated license exports to QWY destinations of technical data related to equipment exports.

The export of technical data under validated license is authorized only to the extent specifically indicated on the face of the license. The only data related to equipment exports that can be provided under general license is the publicly available data authorized by General License *GTDA*, or the assembly, installation, maintenance, repair, and operation data authorized by General License *GTDR*.

The following questions and answers are intended to further clarify this matter as well as related problems involving other general licenses. These are clarifications only. Their issuance in no way affects the suspension of validated licenses for exports to the U.S.S.R. of January 11, 1980 (45 FR 3027).

Question A:

Can we send technicians to repair equipment even though we can no longer supply parts due to suspension or revocation of our validated export license?

Answer:

You can make repairs only to the extent the technical know-how utilized by the technicians does not exceed that contained in the manuals and instructions provided with the equipment under *GTDR*. Under *GTDR*, the technical data must be transmitted within one year of shipment of the equipment. Once the *GTDR* data is transmitted, subsequent service visits can be made at any time and are not restricted to the first year after shipment, but the services performed must be based solely on the data provided under *GTDR*.

Question B:

Must we continue to fulfill visitation and reporting requirements imposed by our suspended (or revoked) validated export licenses even though we can no longer provide parts for repairs we have normally made during these visits?

Answer:

Every effort must be made to continue visitations. If your technicians are not permitted to visit the site or are otherwise prevented from fulfilling your visitation and reporting obligations, you should present evidence of this with your visitation report.

Question C:

We have completed a plant, fully utilizing the licenses we were issued for technical data and equipment. For contractual and warranty reasons, our technicians remain on site. Can they continue to assist and advise the customer even though we no longer have a valid export license?

Answer:

They may advise and assist, but only to the extent permitted by *GTDA* or *GTDR*. If the consignee has been furnished information on installation, operation, maintenance, or repair of equipment, your technicians can repeat that information or perform the needed work, provided they do not exceed the *GTDR* limitations. Your technicians can, for instance, monitor the operation of machines to assure that they are not misused in any way that could result in unfounded warranty claims and possible arbitration. However, your personnel may not perform plant operations or repair functions utilizing technical data requiring a validated license.

Question D:

We have been setting up a turnkey plant. The equipment is there but needs to be installed. Can we install the machines and teach the customer how to run them even though our license has been suspended (or revoked)?

Answer:

Individual machines may be installed and put into operating order, but only to the extent permitted by *GTDR*. Instruction or assistance in combining the individual machines to perform the process for which the plant was intended, however, is not covered by the general license and this service cannot be performed while the requisite validated license for technical data is suspended.

In determining what constitutes an individual machine, exporters should consider as separate machines each of the basic pieces of equipment that join to provide a manufacturing process, but need not separate items such as machine tool/controller combinations or computer mainframe/peripheral systems designed and sold as functional units.

Question E:

If a validated export license is suspended, can a U.S. manufacturer have its field technicians install a product that was exported prior to the suspension?

Answer:

If installation of the product involves no more knowledge than could be absorbed by a competent technician from manuals, instructions, and drawings that normally are delivered with the product under *GTDR* (i.e., necessary for assembly, installation, maintenance, repair, or operation) then the manufacturer's field technicians may install the product. If the field engineers would have to exceed this limitation to install the equipment, then they may not do so. In no case can the field engineers go beyond installation of individual machines to reveal how machines relate to each other to perform a manufacturing process, since process technology is subject to validated license.

Question F:

We applied for a license to export equipment, received a license, and shipped. Subsequently, all validated licenses for that destination were suspended. In our formal application, we described a list of activities that would be required to complete this consolidated business transaction, such as—

—Detailed training schedule for the customer's technical personnel, for maintenance and operation of the equipment;

—Direct support by our engineers for a period of several years to assure proper procedures and use of the equipment;

—A contract for back-up maintenance and repair for equipment, until the customer's engineers

gained sufficient experience to handle the job; and

—A contract for study and analysis and design of their total plant (or a particular department) equipment requirements.

When we got the license for the export of equipment we understood that the license covered the total transaction and permitted the undertaking of the various ancilliary activities as represented in our application.

We have just started these activities, and are obliged under our contract to continue performance even though the license is suspended.

How does the suspension order affect our continued performance, where the equipment is already at the customer's facility waiting for our engineers to begin work and also train the customer's personnel?

Answer:

In the situation you describe, it is not entirely clear how much of your work performance and training is covered by the suspension of your license. It seems likely that a good portion of the work should have been subject to a separate validated technical data license, which would be subject to the general suspension of validated licenses. As you did not approach the transaction (from a licensing procedure viewpoint) in this manner, we will need to review the case to determine what activities are directly related to the provision of technical data under *GTDR* (i.e., required for installation, assembly, maintenance, repair, and operation of the equipment shipped) and which activities go beyond this definition and will require a separate technical data approval for you to continue work.

Question G:

Our engineers sometimes provide technical advice on repair situations by telephone from a Western country to the customer in a country where our validated licenses have been suspended (or revoked). This is particularly likely to occur in difficult situations where the customer will describe his problem in technical detail to our engineer, and our engineer tells him, based on experience and access to more complete equipment history information, where to look for the error. This is an exchange of information, where we also gain knowledge of specific problems with the equipment.

Can this procedure be viewed as *GTDR* information necessary for the repair of installed equipment?

Answer:

This is a reasonable activity that may be covered by *GTDR*, subject to the restriction that it may not involve technical information beyond the scope provided in the manuals and instructions originally given to the customer: *i.e.*, the exchange of information must have as its base of reference the knowledge as provided in those manuals and instructions. The exchange may not, for example, go into detail of component design for engineering of a specific standard part, where the manual does not provide such technical detail.

Provision of such advice may be made at any time, provided it is based solely on *GTDR* information that was furnished within one year of the equipment export.

The above clarification would not only apply to a telephone call from a Western country, but would also be applicable for on-site service calls where the engineer is at the customer location.

Question H:

Our customer installation in a country where our validated licenses have been suspended (or revoked) uses certain standard computer programs shipped some time ago under validated license. The customer normally receives a programming service under warranty-type contract provisions to receive error corrections for these programs. When he has a problem, he identifies it to us (often via a magnetic tape) and a correction (revised program instruction) is sent back to the customer. This exchange is usually mutually advantageous, as we gain from customer experiences and he receives assistance in "fixing" his problem.

Can these exchanges be made under a general license? Can periodic updates (consolidated correction of errors) be provided to the customer?

Answer:

Provided this is routine error correction with no new facilities or performance enhancements provided above the original program specifications, it can be provided either under *GLR* (*e.g.*, a "faulty" tape sent to the U.S. and corrected tape sent back to the customer), or under *GTDR*. These general licenses are available only for one year following export of the equipment, unless extended by the Office of Export Licensing.

Question I:

We are required by our contract to provide for our equipment installed in a country where our validated licenses have been suspended (or revoked), even

after warranty period, engineering changes based on continuing experience with and improvements to that equipment. These changes are for the purpose of, *e.g.*, assuring operation of the machines to the original equipment specifications, removing a particular error-prone condition, or in some cases for the safety of user or maintenance personnel. The transfer of technical data and in some instances new or exchange parts are involved. We consider some of these changes "mandatory" where safety is concerned.

Can we provide the technical data for this purpose under *GTDR* if extended beyond the one year limitation period, and can the requisite parts be provided under *GLR* provisions, where there is no charge to the customer?

Answer:

This technical data may be exported provided it does not exceed the limitations of *GTDR*. However, formal waiver of the one-year limitation must be obtained where such continued shipments are foreseen.

The provision of parts under *GLR* is restricted to warranty situations within one year of export. "Mandatory changes" provided under warranty at no charge are considered an appropriate use of *GLR*. We would consider waiving the one-year limit for export of parts under *GLR* based on the merits of the individual case.

To request extension of the time limit under *GTDR* or *GLR*, the applicant should submit a letter, in duplicate, to the Director, Office of Export Licensing, Export Administration, P.O. Box 273, U.S. Department of Commerce, Washington, D.C. 20044. The letter should specify the export license number, the date of export of the equipment, the reason the extension is needed, whether *GTDR*, *GLR*, or both should be extended, and the period of time for which the extension is desired. Extensions may not exceed one year, but further extensions may be sought if necessary. If approved, OEL will validate the copy of the letter and return it to the applicant.

Question J:

Our engineers, in installing or repairing equipment, use techniques (experience as well as proprietary knowledge of the internal componentry or specifications of the equipment) that exceed what is provided in the standard manuals or instructions (including training) given to the customer. In some cases, it is also a condition of the license that such information provided to the customer be constrained to the minimum necessary for normal installation, maintenance and operation situations.

Can we send an engineer (with knowledge and experience) to the customer site to perform the installation or repair, under the provisions of GTDR, if it is understood that he is restricted by our normal business practices to performing the work without imparting the knowledge or technical data to the customer personnel?

Answer:

As defined in § 379.1, export of technical data includes release of U.S.-origin data in a foreign country, and "release" includes "application to situations abroad of personal knowledge or technical experience acquired in the United States." As the release of technical data in the circumstances described here would exceed that permitted under GTDR, a validated export license would be required even though the technician could apply the data without disclosing it to the customer.

Question K:

We shipped equipment prior to the suspension (or revocation) of our validated licenses, and had planned according to our normal business practices to train customer engineers to maintain that equipment. The training is contractual in nature, provided for a fee, and is scheduled to take place in part in the customer's facility and in part in the U.S. Can we now proceed with this training at both locations under general license authority?

Answer:

Provided that this is your normal training, and involves technical data contained in your manuals and standard instructions for the exported equipment, and meets the other requirements of § 379.4, the training may be provided within the limits of General License GTDR. The location of the training is not significant, as the export occurs at the time and place of the actual transfer or imparting of the technical data to the customer's engineers.

Any training beyond that covered in the GTDR provisions, but specifically represented in your license application as required for this customer installation, and in fact authorized on the face of the license or a separate technical data license, may not be undertaken while the validated license is suspended or revoked.

Question L:

We have contracted with our customer to provide a "customized" service for a fee (*i.e.*, write a program or an instruction manual for a specific customer

requirement), and this was described in detail in our equipment export application. When we received our equipment export license under which we completed shipment and installation, we assumed that this work could commence, being directly related to the equipment and identified by us as necessary for the operation of that equipment for the specific customer.

Can we continue this work, which is already started, or is this impacted by the suspension of our equipment license?

Answer:

In the case you have outlined, a separate validated license is required to cover technical data involved in such "specialized, custom service", unless the equipment validated license clearly specifies on its face that it covers the technical data. In any event, if all validated licenses are suspended to your customer's country the work must cease as it cannot be carried out under general license.

Question M:

In most cases, we ship parts under validated licenses. However, the Regulations do provide for parts shipments for warranty repair situations under the GLR provisions of § 371.17.

If our validated licenses are suspended or revoked, is it permissible to change our previous license procedure and to ship parts on a one-for-one replacement basis where the equipment and parts are under manufacturer's warranty?

Answer:

It is permissible to use any general license (G-DEST, GLR, etc.) that may apply to a given shipment.

Question N:

We have shipped equipment to our customer prior to suspension or revocation of our validated licenses, and have scheduled some management consultation with the customer relating to installation and operational planning, organization, and scheduling of the utilization of the exported equipment. The consultations may include meetings with the customer's management, discussion of specific problems, provision of generally available brochures or basic technical manuals, conducting introductory level training for the operations and support staff, etc.

These consultations, meetings, seminars, training sessions, and manuals are provided consistent with our standard business practices at no additional charge to customers, and are for the purpose of advising all customers of the basic management and organizational techniques involved in utilization of the equipment.

We have previously considered the level of technical data involved as generally available *GTDA* or *GTDR*, without drawing much precise distinction. Do we need a validated license?

Answer:

Generally available published technical data may be provided under *GTDA*. Also, the provision of manuals and instructions beyond the definition of *GTDA* is permissible under *GTDR*, provided that such manuals and instructions are directly required for the installation, assembly, maintenance, repair, or operation of the exported equipment. A seminar, training course, or consultation with the customer based on and restricted to the contents of such *GTDA* or *GTDR* manuals, instructions, and materials would be governed by the provisions of these general licenses.

Also, it is recognized that dialogue with the customer involves transfer of technical data, often of a general nature. In such consultations or dialogues (in the context of training sessions, seminars, and working meetings), it is necessary to limit the content of these exchanges to identifiable published material (*GTDA*), or the manuals and instructions provided together with the equipment under *GTDR*.

Guidance from OEL should be obtained whenever you are in doubt. Where it is determined necessary by OEL, technical data license applications should be filed for situations where technical data ex-

change may exceed that clearly within the restrictions of *GTDR*.

Question O:

We have shipped computer equipment prior to the suspension or revocation of our validated licenses, but have not yet delivered the basic computer control programs (systems control software) required for the operation of the equipment.

Can we ship these programs, and assist the customer in getting them properly installed and operational?

Answer:

Computer software that is not generally available as defined in General License *GTDA* requires a validated license for export, and thus may not be shipped in these circumstances. However, computer software that is both essential for operation of the system and specific to the application for which the system was licensed may be exported under the provisions of *GTDR*. This would include also the manuals or instructions necessary for the installation of such software. Assistance by your personnel in installation or use of *GTDA* or *GTDR* software is limited to the utilization of the materials provided to the customer and within the standard specifications of the software provided.

Additional training or instruction beyond the standard provisions would be subject to validated license.

Updates (error correction or software warranty services) to the original installed software are permissible under *GTDR*, provided that there is no enhancement of performance.

COMMODITIES SUBJECT TO REPUBLIC OF SOUTH AFRICA AND NAMIBIA EMBARGO POLICY

(See § 379.4(e) and § 385.4(a))

(1) Spindle assemblies, consisting of spindles and bearings as a minimal assembly, *except those assemblies with axial and radial axis motion measured along the spindle axis in one revolution of the spindle equal to or greater (coarser) than the following: (a) 0.0008 mm TIR (peak-to-peak) for lathes and turning machines; or (b) D x 2 x 10⁻⁵ mm TIR (peak-to-peak) where D is the spindle diameter in millimeters for milling machines, boring mills, jig grinders, and machining centers* (ECCN 1093);

(2) Equipment for the production of military explosives and solid propellants, as follows:

(a) Complete installations; and

(b) Specialized components (for example, dehydration presses; extrusion presses for the extrusion of small arms, cannon and rocket propellants; cutting machines for the sizing of extruded propellants; sweetie barrels (tumblers) 6 feet and over in diameter and having over 500 pounds product capacity; and continuous mixers for solid propellants) (ECCN 1118);

(3) Specialized machinery, equipment, gear, and specially designed parts and accessories therefor, specially designed for the examination, manufacture, testing, and checking of the arms, appliances, machines, and implements of war (ECCN 2018), ammunition hand-loading equipment for both cartridges and shotgun shells, and equipment specially designed for manufacturing shotgun shells (ECCN 5399).

(4) Construction equipment built to military specifications, specially designed for airborne transport (ECCN No. 2317);

(5) Vehicles specially designed for military purposes, as follows:

(a) Specially designed military vehicles, excluding vehicles listed in Supplement No. 2 to Part 370 (ECCN 2406);

(b) Pneumatic tire casings (*excluding tractor and farm implement types*), of a kind specially constructed to be bulletproof or to run when deflated (ECCN 2406);

(c) Engines for the propulsion of the vehicles enumerated above, specially designed or essentially modified for military use (ECCN 2406); and

(d) Specially designed components and parts to the foregoing (ECCN 2406);

(6) Pressure refuellers, pressure refuelling equipment, and equipment specially designed to facilitate operations in confined areas and ground equipment, not elsewhere specified, developed specially for aircraft and helicopters, and specially designed parts and accessories, n.e.s. (ECCN 2410);

(7) Specifically designed components and parts for ammunition, *except cartridge cases, powder bags, bullets, jackets, cores, shells, projectiles, boosters, fuses and components, primers, and other detonating devices and ammunition belting and linking machines* (ECCN 2603);

(8) Nonmilitary shotguns, barrel length 18 inches or over; and nonmilitary arms, discharge type (for example, stun-guns, shock batons, etc.), *except arms designed solely for signal, flare, or saluting use;* and parts, n.e.s. (ECCN 5998);

(9) Shotgun shells, and parts (ECCN 6998);

(10) Military parachutes (ECCN 2410A);

(11) Submarine and torpedo nets (ECCN 2409A);

(12) Bayonets and muzzle-loading (black powder) firearms (ECCN 2901A).

COMPUTER SOFTWARE

Software described in this Supplement is subject to the written assurance requirements of § 379.4(f)(1)(i)(a) and requires an individual validated license for export to Country Groups QSWYZ, the People's Republic of China. and Afghanistan. Also see § 379.4(e) for written assurance and validated license requirements for exports to South Africa and Namibia.

Technical Notes:

1. "Software" is defined as follows:

"Software"—A collection of one or more "programs" or "microprograms" fixed in any tangible medium of expression.

"Program"—A sequence of instructions to carry out a process in, or convertible into, a form executable by an electronic computer.

"Microprogram"—A sequence of elementary instructions, maintained in a special storage, the execution of which is initiated by the introduction of its reference instruction into an instruction register.

2. "Software" is categorized as follows (there is a close relationship and possible overlap among these categories):

"Development system"—"Software" to develop or produce "software". This includes "software" to manage those activities. Examples of a "development system" are programming support environments, software development environments, and programmer productivity aids.

"Software" to convert a convenient expression of one or more processes ("source code" or "source language") into equipment executable form ("object code" or "object language").

"Diagnostic system"—

"Software" to isolate or direct "software" or equipment malfunctions.

"Maintenance system"—

"Software" to:

(a) Modify "software" or its associated documentation in order to correct faults, or for other updating purposes; or

(b) "Maintain" equipment;

"Operating system"—

"Software" to control:

(a) The operation of a "digital computer" or of "related equipment"; or

(b) The loading or execution of "programs".

"Application software"—

"Software" not falling within any of the definitions of the other categories of "software".

3. "Specifically designed software" is defined as:

The minimum "operating systems", "diagnostic systems", "maintenance systems" and "application software" necessary to be executed on a particular equipment to perform the function for which it was designed. To make other incompatible equipment perform the same function requires:

(a) Modification of this "software"; or

(b) Addition of "programs".

(For a complete list of definitions of terms used in this Supplement, see Advisory Note 12 below; see also ECCN 1565 for additional definitions relating to electronic computers.)

List of Software Subject To This Supplement to Part 379

(a) "Software" of whatever category, as follows:

(1) "Software" designed or modified for any computer that is part of a computer series designed and produced within Country Groups, Q, W, Y, or Z, the People's Republic of China, or Afghanistan: except "application software" designed for and limited to:

(i) Accounting, general ledger, inventory control, payroll, accounts receivable, personnel records, wages calculation or invoice control;

(ii) Data and text manipulation such as sort/merge, text editing, data entry or word processing;

(iii) Data retrieval from established data files for purposes of report generation or inquiry for the functions described in (i) or (ii) above; or

(iv) The non "real time processing" of pollution sensor data at fixed sites or in civil vehicles for civil environmental monitoring purposes;

(2) "Software" designed or modified for the design, development or production of items controlled by ECCNs on the Commodity Control List identified by the code letter "A", by the International Traffic in Arms Regulations, by 10 CFR 110 or by 10 CFR 810;

(3) "Software" designed or modified for:

(i) Controlled "hybrid computers";

(ii) One or more of the functions described in ECCN 1565(h)(1)(i)(A) to (J) and (M) or for "digital computers" or "related equipment" designed or modified for such functions, *except* the minimum "specially designed software" in machine executable form for "digital computers" and "related equipment" therefor that are freed from controls only by ECCN 1565(h)(2)(i) or (ii), and only when supplied with the equipment or systems;

(4) "Software" for computer-aided design, manufacture, inspection or test of items controlled by ECCNs on the Commodity Control List identified by the code letter "A", by the International Traffic in Arms Regulations, by 10 CFR 110 or by 10 CFR 810;

(5) "Software" designed or modified to provide certifiable multi-level security or certifiable user isolation applicable to government classified material or to applications requiring an equivalent level of security, or "software" to certify such "software"[1];

(b) Categorized "software", as follows:

---

[1]Department of Defense certifiable under section (b×3) of "Department of Defense Trusted Computer System Evaluation Criteria", published in DOD Computer Security Center, Fort Meade, MD 20755.

(1) "Development systems":

(i) "High-level language" "development systems" designed for or containing "programs" or "databases" special to the development or production of:

(A) "Specially designed software" controlled by ECCNs on the Commodity Control List identified by the code letter "A", by the International Traffic in Arms Regulations, by 10 CFR 110 or by 10 CFR 810;

(B) "Software" controlled by sub-paragraphs (a)(2) or (a)(3) of this Supplement, including any subset designed or modified for use as part of such a "development system";

(ii) "High-level language" "development systems" designed for, or containing the "software" tools and "databases" for, the development or production of "software", or any subset designed or modified for use as part of a "development system" such as or equivalent to:

(A) Ada Programming Support Environment (APSE);

(B) Any subset of APSE, as follows:

(1) Kernel APSE;

(2) Minimal APSE;

(3) Ada compilers specially designed as an integrated subset of APSE; or

(4) Any other subset of APSE;

(C) Any superset of APSE; or

(D) Any derivative of APSE;

(2) "Programming systems";

(i) "Cross-hosted" compilers and "cross-hosted" assemblers;

Note: For "cross-hosted" compilers or "cross-hosted" assemblers that have to be used in conjunction with microprocessor or microcomputer development instruments or systems described in ECCN 1529A, see that ECCN.

(ii) Compilers or interpreters designed or modified for use as part of a "development system" controlled by subparagraph (b)(1) above;

(iii) Disassemblers, decompilers or other "software" that convert "programs" in object or assembly language into a higher level language, except simple debugging "application software", such as mapping, tracing, checkpoint/restart, breakpoint, dumping and the display of the storage contents or their assembly language equivalent;

(3) "Diagnostic systems" or "maintenance systems" designed or modified for use as part of a "development system" controlled by sub-paragraph (b)(i) above;

(4) "Operating systems":

(i) "Operating systems" designed or modified for "digital computers" or "related equipment" exceeding any of the following limits:

(A) Central processing unit—"main storage" combinations:

(1) "Total processing data rate"—48 million bit per second;

(2) "Total connected capacity" of "main storage"—25.2 million bit;

(3) "Virtual storage" capability—512 M Byte;

(B) Input-output control unit-drum, disk or cartridge-type streamer tape drive combinations:

(1) "Total transfer rate"—15 million bit per second;

(2) "Total access rate"—320 access per second;

(3) Total connected "net capacity"—7,000 million bit;

(4) "Maximum bit transfer rate" of any drum or disk drive—10.3 million bit per second;

(C) Input/output control unit-bubble memory combinations: Total connected "net capacity"—2.1 million bit;

(D) Input/output control unit—magnetic tape drive combinations:

(1) "Total transfer rate"—5.2 million bit per second;

(2) Number of magnetic tape drives—twelve;

(3) "Maximum bit transfer rate" of any magnetic tape drive—2.6 million bit per second;

(4) "Maximum bit packing density"—63 bit per mm. (1,600 bit per inch) per track;

(5) Maximum tape read/write speed—508 cm. (200 inch) per second;

Note: This sub-paragraph does not control "operating systems" designed or modified for "digital computers" or "related equipment":

(a) Not exceeding the above limits even when the "operating systems" can also be used on "digital computers" or "related equipment" exceeding the above limits; or

(b) Belonging to a series containing models exceeding the above limits, if the "operating systems" are used on "digital computers" or "related equipment" of the series that do not exceed the above limits.

(ii) "Operating systems" providing online transaction data processing that permit integrated teleprocessing and "on-line updating" of "databases";

(5) "Application software";

(i) "Software" for cryptologic or cryptanalytic applications;

(ii) Artificial intelligence "software", including "software" normally classified as expert systems, that enables a "digital computer" to perform functions normally associated with human perception and reasoning or learning;

(iii) "Database management systems" designed to handle "distributed databases" for

(A) Fault tolerance by using techniques such as maintenance of duplicated "databases"; or

(B) Integrating data at a single site from independent remote "databases";

(iv) "Software" designed to adapt "software" resident on one "digital computer" for use on another "digital computer", except "software" to adapt between two legally exported machines.

(c) "Specially designed software" for equipment, as follows:

(1) Navigation and direction finding equipment controlled under ECCN 1501A(b);

(2) Radar equipment controlled under ECCN 1501A(c);

(3) Equipment controlled under ECCN 1502A;

(4) Equipment controlled under ECCN 1510A;

(5) Equipment controlled under ECCN 1516A;

(6) Equipment controlled under ECCN 1519A;

(7) Equipment controlled under ECCN 1520A;

(8) Equipment controlled under ECCN 1567A;

(9) Equipment controlled under ECCN 1529A;

(10) Equipment controlled under ECCN 1533A;

(d) "Specially designed software" for the following:

(1) Technical data described in Supplement No. 4 to Part 379 for metal-working manufacturing processes;

(2) Spin-forming and flow-forming machines controlled under ECCN 1075A;

(3) Equipment, tooling and fixtures controlled under ECCN 1080A for the manufacturing or measuring of gas turbine blades or vanes;

(4) Equipment, tools, dies, molds and fixtures controlled under ECCN 1081A for the manufacture or inspection of aircraft, airframe structures, or aircraft fasteners;

(5) Equipment, tools, dies, molds, fixtures and gauges controlled under ECCN 1086A for the manufacture or inspection of aircraft and aircraft-derived gas turbine engines;

(6) Electric vacuum furnaces controlled under ECCN 1203A;

(7) Electric arc devices controlled under ECCN 1206A;

(8) Metal rolling mills controlled under ECCN 1305A;

(9) Isostatic presses controlled under ECCN 1312A;

(10) Equipment controlled under ECCN 1354A for the manufacture or testing of printed circuit boards;

(11) Equipment controlled under ECCN 1357A for the production of fibers controlled by ECCN 1763A, or their composites;

(12) Equipment controlled under ECCN 1358A for the manufacture or testing of devices and assemblies controlled under ECCN 1588A and ECCN 1572A;

(13) Test facilities and equipment controlled under ECCN 1361A for the design or development of aircraft or gas turbine aero-engines;

(14) Water tunnel equipment controlled under ECCN 1363A, including software that contains databases generated by the use of equipment controlled under that ECCN;

(15) Equipment controlled under ECCN 1365A specially designed for in-service monitoring of acoustic emissions in airborne vehicles or underwater vehicles;

(16) Machine tools controlled under ECCN 1370A for generating optical quality surfaces;

(17) Computer-controlled pumping and flooding systems that will permit the docking of listing vessels when used with floating docks controlled under ECCN 1425A;

(18) Design and production of aircraft and helicopter airframes and propulsion systems controlled under ECCN 1460A;

(19) Integration software for integrated flight instrument systems, automatic pilots, inertial or other equipment using accelerometers or gyros controlled under ECCN 1485A;

(20) Marine or terrestrial acoustic or ultrasonic systems or equipment controlled under ECCN 1510A; and

(21) Precision linear and angular measuring systems controlled under ECCN 1532A.

Advisory Notes

Advisory Note 1: Reserved.

Advisory Note 2: Reserved.

Advisory Note 3: Licenses are likely to be approved for export to satisfactory end-users in Country Groups QWY, the People's Republic of China (PRC) and Afghanistan of "software" initially exported to those destinations before January 1, 1984, provided that:

(a) The "software" is identical to and in the same language form (source or object) as initially exported, allowing minor updates for the correction of errors that do not modify the initially exported functions;

(b) The accompanying documentation does not exceed the level of the initial export;

(c) The "software" is exported to the same controlled destination as the initial export.

Advisory Note 4: Licenses are likely to be approved for export to satisfactory end-users in Country Groups QWY, the People's Republic of China (PRC) and Afghanistan of "application software" controlled by sub-paragraph (a)(1) above, but not otherwise listed in this Supplement or ECCNs on the Commodity Control List identified by the code letter "A", provided that:

(a) The "application software" is designed for and limited to the following:

(1) The approved end-use of legally exported equipment or systems in conjunction with any computer that is part of a computer series produced within a controlled area and based on a design originating in a COCOM country; or

(2) The monitoring and control of industrial processes limited to the production of items not described by ECCNs on the Commodity Control List identified by the code letter "A", by the International Traffic in Arms Regulations, by 10 CFR 110 or by 10 CFR 810; and

(b) No restricted technical data is provided.

Advisory Note 5: Licenses are likely to be approved for export to satisfactory end-users in Country Groups QWY, the People's Republic of China (PRC) and Afghanistan of "software" not exceeding 5,000 statements in "source language", excluding data, provided that:

(a) The "software" is neither designed nor modified for use as a module of a larger "software" module or system that in total exceeds this limit;

(b) The "software" is not controlled by sub-paragraph (b)(5) above; and

(c) The Office of Export Licensing is reasonably satisfied that:

(1) The "software" will be used primarily for the specific non-strategic application for which the export would be approved;

(2) The type and characteristics of such "software" are reasonable for this application; and

(3) The "software" will not be used for the design, development or production of items controlled by ECCNs on the Commodity Control List identified by the code letter "A", by the International Traffic in Arms Regulations, by 10 CFR 110 or by 10 CFR 810.

Advisory Note 6: Reserved.

Advisory Note 7: Reserved.

Advisory Note 8: Licenses are likely to be approved for export to satisfactory end-users in Country Groups QWY, the People's Republic of China (PRC) and Afghanistan of normal commercial "software" for civil Air Traffic Control (ATC) systems approved for export, provided that:

(a) The "software" is commonly used by civil Air Traffic Control authorities outside controlled areas, but not precluding the personalization of certain parameters for civil Air Traffic Control authorities wherever located:

(b) The "software" is not designed or modified for any "digital computer" that is part of a "digital computer" series designed and produced within a controlled area;

(c) The "software" is the minimum necessary to accomplish the normal civil Air Traffic Control functions outside controlled areas;

(d) The "software" will not contain or be capable of accomplishing any of the following functions:

(1) Electronic Counter Counter Measures (ECCM);

(2) Weapon display, allocation or operation;

(3) Intercept guiding capability; or

(4) Interfacing with altitude determining radars, except secondary search radars;

(e) The "software" is further limited by the amount of "source code", which is to be the minimum necessary for the use (i.e., installation, operation and maintenance) of the "software";

(f) In addition to the above limitations, the only other system "software" allowed is the minimum "programming system" for the maintenance of the "software";

(g) A signed statement of the end-user or importing agency containing a full description of the "software" and its characteristics vis-a-vis the sub-paragraph above, its intended application and workload and a complete identification of all end-users and their activities is provided;

(h) The "software" will not be used to provide or process data associated with military control centers or military radars or otherwise be associated with such radars or centers; and

(i) The type and characteristics of the "software" are reasonable for the specific civil Air Traffic Control applications.

Advisory Note 9: Licenses are likely to be approved for export to satisfactory end-users in Country Group QWY, the People's Republic of China (PRC) and Afghanistan of "operating systems" controlled only by sub-paragraph (b)(4)(ii) above when supplied with "digital computer" and "related equipment" exported under the provisions of ECCN 1565, Advisory Notes 9 and 12, provided that these "operating systems" are:

(a) For use with a "digital computer" exported under the provisions of ECCN 1565;

(b) In machine executable version;

(c) Limited to the minimum "standard commercially available" "software"; and

(d) Not designed or modified for "database management systems" controlled by sub-paragraph (b)(5)(iii) above.

Advisory Note 10: Licenses are likely to be approved for export to satisfactory end-users in Country Group QWY, the People's Republic of China (PRC) and Afghanistan of "software" controlled by sub-paragraph (a)(3)(ii) above the "digital computers" and "related equipment" exported under the provisions of ECCN

1529, Advisory Note 5, or ECCN 1565, Advisory Notes 5 and 9, provided that:

(a) The "software" is limited to:

(1) The minimum necessary for the approved application:

(2) Machine executable form; and

(3) "Specially designed software" for

(i) Equipment likely to be approved for export solely under ECCN 1529, Advisory Note 3;

(ii) Equipment likely to be approved for export under ECCN 1565, Advisory Note 5, for one or more of the functions described in ECCN 1565(h)(1)(i)(A), (B), or (D);

(iii) Equipment likely to be approved for export under ECCN 1565, Advisory Note 9, for one or more of the functions described in ECCN 1565(h)(1)(i)(A), (B), or (C);

(b) The "specially designed software" for "signal processing" and "image enhancement" does not provide for more than one of the following:

(1) Time compression; or

(2) Transformations between domains (e.g. Fast Fourier Transform or Walsh Transform).

Advisory Note 11: Licenses are likely to be approved for export to satisfactory end-users in Country Groups QWY, the People's Republic of China (PRC) and Afghanistan of "software" controlled by sub-paragraph (a)(3)(ii) above for "digital computers" and "related equipment" exported under the provisions of ECCN 1565, Advisory Note 12, provided that the "software" is limited to

(a) "Software" for one or more of the functions described in ECCN 1565(h)(1)(i)(A), (B) or (C);

(b) The minimum necessary for the approved applications; and

(c) Machine executable form.

Advisory Note 12: Definitions of Terms Used in this Supplement:

"Analog computer"—

Equipment that can, in the form of one or more continuous variables:

(a) Accept data;

(b) Process data; and

(c) Provide output of data.

"Application software"—

"Software" not falling within any of the definitions of the other categories of "software".

"Cross-hosted"—

For "programming systems", those that produce "programs" for a model of electronic computer different from that used to run the "programming system", i.e., they have code generators for equipment different from the host computer.

"Database"—

A collection of data, defined for one or more particular applications, physically located and maintained in one or more electronic computers or "related equipment".

"Database management system"—

"Applications software" to manage and maintain a "database" in one or more prescribed logical structures for use by other "application software" independent of the specific methods used to store or retrieve the "database".

"Development system"—

"Software" to develop or produce "software". This includes "software" to manage those activities. Examples of a "development system" are programming support environments, software development environments, and programming productivity aids.

"Diagnostic system"—

"Software" to isolate or detect "software" or equipment malfunctions.

"Digital computer"—

Equipment that can, in the form of one or more discrete variables:

(a) Accept data;

(b) Store data or instructions in fixed or alterable (writable) storage devices;

(c) Process data by means of stored sequence of instructions that is modifiable; and

(d) Provide output of data.

Note: Modifications of a stored sequence of instructions include replacement of fixed storage devices, but not a physical change in wiring or interconnections.

"Distributed database"—A "database" physically located and maintained in part or as a whole in two or more interconnected electronic computers or "related equipment", such that inquiries from one location can involve "database" access in other interconnected electronic computers or "related equipment".

"Firmware"—see "microprogram".

"High-level language"—A programming language that does not reflect the structure of any one given electronic computer or that of any one given class of electronic computers.

"Hybrid computer"—Equipment that can:

(a) Access data;

(b) Process data, in both analog and digital representations; and

(c) Provide output of data.

"Maintenance system"—"Software" to:

(a) Modify "software" or its associated documentation in order to correct faults, or for other updating purposes; or

(b) Maintain equipment.

"Microprogram"—A sequence of elementary instructions, maintained in a special storage, the execution of which is initiated by the introduction of its reference instruction into an instruction register.

"Object code" or *"object language"*—See "programming system".

"On-line updating"—Processing in which the contents of a "database" can be amended within a period of time useful to interact with an external request.

"Operating system"—"Software" to control:

(a) The operation of a "digital computer" or of "related equipment"; or

(b) The loading or execution of "programs".

"Program"—A sequence of instructions to carry out a process in, or convertible into, a form executable by an electronic computer.

"Programming system"—"Software" to convert a convenient expression of one or more processes ("source code" or "source language") into equipment executable form ("object code" or "object language").

"Related equipment"—Equipment 'embedded' in, 'incorporated' in, or 'associated' with electronic computers, as follows:

(a) Equipment for interconnecting "analog computers" with "digital computers";

(b) Equipment for interconnecting "digital computers";

(c) Equipment interfacing electronic computers to "local area networks" or to "wide area networks";

(d) Communication control units;

(e) Other input/output (I/O) control units;

(f) Recording or reproducing equipment referred to ECCN 1565 by ECCN 1572;

(g) Displays; or

(h) Other peripheral equipment.

Note: "Related equipment" containing an "embedded" or "incorporated" electronic computer, but lacking "user accessible programmability", does not thereby fall within the definition of an electronic computer.

"Self-hosted"—For "programming systems", those producing "programs" for the same model of electronic computer as that used to run the "programming system", i.e., they only have code generators for the host computer.

"Software"—A collection of one or more "programs" or "microprograms" fixed in any tangible medium of expression.

"Source code" or *"source language"*—See "programming system".

"Specially designed software"—The minimum "operating systems", "diagnostic systems", "maintenance systems" and "application software" necessary to be executed on a particular equipment to perform the function for which it was designed. To make other incompatible equipment perform the same function requires:

(a) Modification of this "software" or

(b) Addition of "programs".

"Standard commercially available"—For "software" that is

(a) Commonly supplied to general purchasers or users of equipment outside controlled areas, but not precluding the personalization of certain parameters for individual customers wherever located;

(b) Designed and produced for civil applications;

(c) Not designed or modified for any "digital computer" that is part of a "digital computer" series designed and produced within a controlled area; and

(d) Supplied in a commonly distributed form.

Technical Note: In the case of "software" for mainframe "digital computers" that may have a "virtual storage capability" exceeding the limit of sub-paragraph (b)(4)(i)(A)(3) and that may be considered for export under the conditions of ECCN 1565, Advisory Notes 9 & 12, the limitation of the "virtual storage capability" of 512 MByte does not apply.

Advisory Note for the People's Republic of China: Licenses are likely to be aproved for export to satisfactory end-users in the

People's Republic of China of "software" controlled for export in this supplement No. 3 to Part 379, as follows:

(a) "Software" controlled only by sub-paragraph (a)(1) (of the "List of Software Subject to this Supplement to Part 379") for computers designed and produced within the People's Republic of China;

(b) "Software" controlled by sub-paragraph (a)(3)(ii) for equipment that is covered by an Advisory Note to ECCN 1565A in the Commodity Control List;

(c) "Software" not specially designed for computer-aided design, manufacture, inspection or testing of products controlled for export on the Commodity Control List;

(d) "Cross-hosted" compilers or "cross-hosted" assemblers controlled by sub-paragraph (b)(2)(i);

(e) "Software" controlled by sub-paragraphs (b)(2)(ii) or (b)

(3) for microprocessor or microcomputer development systems

that are covered by an Advisory Note on the Commodity Control List;

(f) "Operating systems" controlled by sub-paragraph (b)(4) for computers that are covered by an Advisory Note to ECCN 1565A on the Commodity Control List.

ADDITIONAL SPECIFICATIONS FOR CERTAIN TECHNICAL DATA REQUIRING A VALIDATED LICENSE TO ALL DESTINATIONS EXCEPT CANADA

(1) Technical data for metal-working manufacturing processes and specially designed software therefor (§ 379.4(d)(16)):

(i) The following are definitions of terms used in section (1) of this Supplement:

(A) "Hot die forging" is a deformation process where die temperatures are at the same nominal temperature as the workpiece and exceed 850 K (577°C, 1,070°F);

(B) "Superplastic forming" is a deformation process using heat for metals that are normally characterized by low values of elongation (less than 20%) at the breaking point as determined at room temperature by conventional tensile strength testing, in order to achieve elongations during processing that are at least 2 times those values;

(C) "Diffusion bonding" is a solid-state molecular joining of at least two separate metals into a single piece with a joint strength equivalent to that of the weakest material;

(D) "Metal powder compaction" is a process capable of yielding parts having a density of 98% or more of the theoretical maximum density;

(E) "Direct-acting hydraulic pressing" is a deformation process that uses a fluid-filled flexible bladder in direct contact with the workpiece;

. (F) "Hot isostatic densification" is a process of pressurizing a casting at temperatures exceeding 375 K (102°C, 215.6°F) in a closed cavity through various media (gas, liquid, solid particles, etc.) to create equal force in all directions to reduce or eliminate internal voids in the casting;

(G) "Vacuum hot pressing" is a process that uses a press with heated dies to consolidate metal powder under reduced atmospheric pressure into a part;

(H) "High pressure extrusion" is a process yielding a single-pass reduction ratio of 4 to 1 or greater in a cross sectional area of the resulting part;

(I) "Isostatic pressing" is a process that uses a pressurizing medium (gas, liquid, solid particles, etc.) in a closed cavity to create equal force in all directions upon a metal powder-filled container for consolidating the powder into a part.

(ii) Technical data covered by paragraph (d)(16) of § 379.4 is as follows:

(A) Technical data for the design of tools, dies and fixtures specially designed for the following processes:

(1) "Hot die forging";

(2) "Superplastic forming";

(3) "Diffusion bonding";

(4) "Metal powder compaction" using:

(i) "Vacuum hot pressing";

(ii) "High-pressure extrusion"; or

(iii) "Isostatic pressing";

(5) "Direct-acting hydraulic pressing";

(B) Technical data consisting of process parameters as listed below used to control:

(1) "Hot die forging":

(i) Temperature;

(ii) Strain rate;

(2) "Superplastic forming" of aluminum alloys, titanium alloys and superalloys:

(i) Surface preparation;

(ii) Strain rate;

(iii) Temperature;

(iv) Pressure;

(3) "Diffusion bonding" of superalloys and titanium alloys:

(i) Surface preparation;

(ii) Temperature;

(iii) Pressure;

(4) "Metal powder compaction" using:

(i) "Vacuum hot pressing":

(a) Temperature;

(b) Pressure;

(c) Cycle time;

(ii) "High-pressure extrusion":

(a) Temperature;

(b) Pressure;

(c) Cycle time;

(iii) "Isostatic pressing":

(a) Temperature;

(b) Pressure;

(c) Cycle time;

(5) "Direct-acting hydraulic pressing" of aluminum alloys and titanium alloys:

(i) Pressure;

(ii) Cycle time;

(6) "Hot isostatic densification" of titanium alloys, aluminum alloys and superalloys:

(i) Temperature;

(ii) Pressure;

(iii) Cycle time.

(2) "Software" and technical data for "automatically controlled industrial systems" to produce assemblies or discrete parts (§ 379.4(d)(17)):

(a) "Software" with all the following characteristics:

(1) Specially designed for "automatically controlled industrial systems" that include at least eight pieces of the equipment enumerated in Technical Note (b)(1) to (9) below;

Notes: 1. The "digital computers" of the "automatically controlled industrial system" do not share a common "main storage" but exchange information by transmitting messages through a "local area network".

2. This sub-paragraph (a)(1) does not release from export control "software" in source code.

(2) Integrating, in a hierarchical manner, while having access to data that may be stored outside the supervisory "digital computer", the manufacturing processes with:

(i) Design functions; or

(ii) Planning and scheduling functions; and

(3) (i) Automatically generating and verifying the manufacturing data and instructions, including selection of equipment and sequences of manufacturing operations, for the manufacturing process from design and manufacturing data; or

(ii) Automatically reconfiguring the "automatically controlled industrial system" through reselecting equipment and sequences of manufacturing operation by "real-time processing" of data pertaining to anticipated but unscheduled events; and

Note: This sub-paragraph (3)(ii) does not control "software" that only provides rescheduling of functionally identical equipment within "flexible manufacturing units" using prestored "part" programs and a prestored strategy for the distribution of the "part" programs.

(b) Technical data for the design of "automatically controlled industrial systems" that will be used with the "software" controlled for export by sub-paragraph (a) above, regardless of whether the conditions of sub-paragraph (a)(1) are met.

Technical Note: For the purposes of paragraph (d)(17) of § 379.4:

(a) An "automatically controlled industrial system" is a combination of:

(1) One or more "flexible manufacturing units"; and

(2) A supervisory "digital computer" for coordination of the independent sequences of computers instructions to, from, and within the "flexible manufacturing units";

(b) A "flexible manufacturing unit" is an entity that comprises a combination of a "digital computer" including its own "main storage" and its own "related equipment", and at least one of the following:

(1) A machine tool or a dimensional inspection machine covered by ECCNs 1091A or 1370A;

(2) A "robot" covered by ECCN 1391A;

(3) A digitally controlled spin-forming or flow-forming machine covered by ECCN 1075A;

(4) Digitally controlled equipment covered by ECCNs 1080A, 1081A, 1086A or 1088A;

(5) Digitally controlled electric arc device covered by ECCN 1206A;

(6) Digitally controlled equipment covered by ECCN 1354A or by paragraph (b) of ECCN 1355A;

(7) Digitally controlled equipment covered by ECCN 1357A;

(8) Digitally controlled electronic equipment covered by ECCN 1529A; or

(9) A digitally controlled measuring system covered by ECCN 1532A.

Note: For the definitions of other terms in quotation marks, see ECCNs 1391A or 1565A or Supplement No. 3 to Part 379.

Note: Sub-paragraph (a) above does not control "software" (in "machine executable form" only) for industrial sectors other than nuclear, aerospace, shipbuilding, heavy vehicles, machine building, microelectronics and electronics. This Note does not release from export control design technology specified in sub-paragraph (b) above.

(3) Technical data for application to non-electrical devices to achieve:

(i) Inorganic overlay coatings or inorganic surface modification coatings,

(A) Specified in column 3 of the Table below,

(B) On substrates specified in column 2 of the Table below,

(C) By processes as defined in **Technical Note** (a) to (h) and specified in column 1 of the Table below, and specially designed software therefor (§ 379.4(d)(19):

TABLE

[This Table should be read to control the technology of a particular coating process only when the resultant coating in column 3 is in a paragraph directly across from the relevant substrate under column 2.

For example, chemical vapor deposition coating process technical data *are controlled* for the application of noble metal modified aluminides to superalloy substrates, but are *not controlled* for the application of noble metal modified aluminides to titanium alloys. In the second case, the resultant coating is *not* listed in the paragraph under column 3 directly across from the paragraph under column 2 listing "Titanium alloys".]

1. Coating Process[1]	2. Substrate	3. Resultant Coating
A. "Chemical Vapor Deposition" (CVD)	Superalloys	Aluminides for internal surfaces, Alloyed aluminides[2], or Noble metal modified aluminides[3]
	Titanium or Titanium alloys	Carbides, Aluminides, or Alloyed aluminides[2]
	Ceramics	Silicides or Carbides
	Carbon-carbon, Carbon-ceramic, or Metal matrix composites	Silicides, Carbides, Mixtures thereof[4], or Dielectric layers
	Copper or Copper alloys	Tungsten or Dielectric layers
	Silicon carbide or Cemented tungsten carbide	Carbides, Tungsten, Mixtures thereof[4], or Dielectric layers
B. "Electron-Beam Physical Vapor Deposition" (EB-PVD)	Superalloys	Alloyed silicides, Alloyed aluminides[2], MCrAlX (*except* CoCrAlY containing less than 22 weight percent of chromium and less than 12 weight percent of aluminum and less than 2 weight percent of yttrium)[5], Modified zirconia (*except* calcia-stabilized zirconia), or Mixtures thereof (including mixtures of the above with silicides or aluminides)[4]
	Ceramics	Silicides or modified zirconia (*except* calcia-stabilized zirconia)
	Aluminum alloys[6]	MCrAlX (*except* CoCrAlY containing less than 22 weight percent of chromium and less than 12 weight percent of aluminum and less than 2 weight percent of yttrium)[5], modified zirconia (*except* calcia-stabilized zirconia) or mixtures thereof[4]
	Corrosion resistant steel[7]	MCrAlX (*except* CoCrAlY containing less than 22 weight percent of chromium and less than 12 weight percent of aluminum and less than 2 weight percent of yttrium)[5] or modified zirconia (*except* calcia-stabilized zirconia)
	Carbon-carbon, Carbon-ceramic, or Metal matrix composites	Silicides, Carbides, Mixtures thereof[4], or Dielectric layers
	Copper or Copper alloys	Tungsten or Dielectric layers
	Silicon carbide or Cemented tungsten carbide	Carbides, Tungsten, Mixtures thereof[4], or Dielectric layers
C. "Electrophoretic deposition"	Superalloys	Alloyed aluminides[2] or Noble metal modified aluminides[3]

1. Coating Process[1]	2. Substrate	3. Resultant Coating
D. "Pack cementation" (see also A above)[9]	Superalloys	Alloyed aluminides[2] or Noble metal modified aluminides[3]
	Carbon-carbon, Carbon-ceramic, or Metal matrix composites	Silicides, Carbides, or Mixtures thereof[4]
	Aluminum alloys[6]	Aluminides or alloyed aluminides[2]
E. "Plasma spraying" (high velocity or low pressure only)	Superalloys	MCrAlX (*except* CoCrAlY containing less than 22 weight percent of chromium and less than 12 weight percent of aluminum and less than 2 weight percent of yttrium)[5], Modified zirconia (*except* calcia-stabilized zirconia), or Mixtures thereof[4]
	Aluminum alloys[6]	MCrAlX (*except* CoCrAlY containing less than 22 weight percent of chromium and less than 12 weight percent of aluminum and less than 2 weight percent of yttrium)[5], Modified zirconia (*except* calcia-stabilized zirconia), Silicides, or Mixtures thereof[4]
	Corrosion resistant steel[7]	MCrAlX (*except* CoCrAlY containing less than 22 weight percent of chromium and less than 12 weight percent of aluminum and less than 2 weight percent of yttrium)[5], Modified zirconia (*except* calcia-stabilized zirconia), or Mixtures thereof[4]
	Titanium or Titanium alloys	Carbides or Oxides
F. "Slurry deposition"	Refractory metals[8]	Fused silicides or Fused aluminides
	Carbon-carbon, Carbon-ceramic, or Metal matrix composites	Silicides, Carbides, or Mixtures thereof[4]
G. "Sputtering" (high rate, reactive, or radio frequency only)	Superalloys	Alloyed silicides, Alloyed aluminides[2], Noble metal modified aluminides[3], MCrAlX (*except* CoCrAlY containing less than 22 weight percent of chromium and less than 12 weight percent of aluminum and less than 2 weight percent of yttrium)[5], Modified zirconia (*except* calcia-stabilized zirconia), Platinum, or Mixtures thereof (including mixtures of the above with silicides or aluminides)[4]
	Ceramics	Silicides, Platinum, or Mixtures thereof[4]
	Aluminum alloys[6]	MCrAlX (*except* CoCrAlY containing less than 22 weight percent of chromium and less than 12 weight percent of aluminum and less than 2 weight percent of yttrium)[5], Modified zirconia (*except* calcia-stabilized zirconia), or Mixtures thereof[4]
	Corrosion resistant steel[7]	MCrAlX (*except* CoCrAlY containing less than 22 weight percent of chromium and less than 12 weight percent of aluminum and less than 2 weight percent of yttrium)[5], Modified zirconia (*except* calcia-stabilized zirconia), or Mixtures thereof[4]
	Titanium or Titanium alloys	Borides or Nitrides
	Carbon-carbon, Carbon-ceramic, or Metal matrix composites	Silicides, Carbides, Mixtures thereof[4], or Dielectric layers
	Copper or Copper alloys	Tungsten or Dielectric layers
	Silicon carbide or Cemented tungsten carbide	Carbides, Tungsten, or Dielectric layers
H. "Ion implantation"	High temperature bearing steels	Additions of chromium, tantalum, or niobium (columbium)
	Beryllium or Beryllium alloys	Borides

1. Coating Process[1]	2. Substrate	3. Resultant Coating
	Carbon-carbon, Carbon-ceramic, or Metal matrix composites	Silicides, Carbides, Mixtures thereof[4], or Dielectric layers
	Titanium or Titanium alloys	Borides or Nitrides
	Silicon nitride or Cemented tungsten carbide	Nitrides, Carbides, or Dielectric layers
	Sensor window materials transparent to electromagnetic waves, as follows: silica, alumina, silicon, germanium, zinc sulphide, zinc selenide, or gallium arsenide	Dielectric layers

Footnotes:

1 Coating process includes coating repair and refurbishing as well as original coating.

2 Multiple-stage coatings in which an element or elements are deposited prior to application of the aluminide coating, even if these elements are deposited by another coating process, are included in the term "alloyed aluminide" coating, but the multiple use of single-stage "pack cementation" processes to achieve alloyed aluminides is not included in the term "alloyed aluminide" coating.

3 Multiple-stage coatings in which the noble metal or noble metals are laid down by some other coating process prior to application of the aluminide coating are included in the term "noble metal modified aluminide" coating.

4 Mixtures consist of infiltrated material, graded compositions, co-deposits and multilayer deposits and are obtained by one or more of the coating processes specified in this Table.

5 MCrAlX refers to an alloy where M equals cobalt, iron, nickel or combinations thereof, and X equals hafnium, yttrium, silicon or other minor additions in various proportions and combinations.

6 Aluminum alloys as a substrate in this Table refers to alloys usable at temperatures above 500K (227°C).

7 Corrosion resistant steel refers to AISI (American Iron and Steel Institute) 300 series or equivalent national standard steels.

8 Refractory metals as a substrate in this Table consist of the following metals and their alloys: niobium (columbium), molybdenum, tungsten, and tantalum.

9 This does not control technical data for single-stage "pack cementation" of solid airfoils.

Technical Note: The definitions of processes specified in column 1 of the Table are as follows:

(a) "Chemical Vapor Deposition" (CVD) is an overlay coating or surface modification coating process wherein a metal, alloy, composite, or ceramic is deposited upon a heated substrate. Gaseous reactants are reduced or combined in the vicinity of a substrate resulting in the deposition of the desired elemental, alloyed, or compounded material on the substrate. Energy for this decomposition or chemical reaction process is provided by the heat of the substrate.

Note 1: CVD includes the following processes: out-of-"pack", pulsating, controlled nucleation thermal decomposition (CNTD), plasma enhanced or plasma assisted.

Note 2: "Pack" denotes a substrate immersed in a powder mixture.

Note 3: The gaseous material utilized in the out-of-"pack" process is produced using the same basic reactions and parameters as the "pack cementation" process, except that the substrate to be coated is not in contact with the powder mixture.

(b) "Electron-Beam Physical Vapor Deposition" (EB-PVD) is an overlay coating process conducted in a vacuum chamber, wherein an electron beam is directed onto the surface of a coating material causing vaporization of the material and resulting in condensation of the resultant vapors onto a substrate positioned appropriately.

Note: The addition of gases to the chamber during the processing is an ordinary modification to the process.

(c) "Electrophoretic deposition" is a surface modification coating or overlay coating process in which finely divided particles of a coating material suspended in a liquid dielectric medium migrate under the influence of an electrostatic field and are deposited on an electrically conducting substrate.

Note: Heat treatment of parts after coating materials have been deposited on the substrate, in order to obtain the desired coating, is an essential step in the process.

(d) "Pack cementation" is a surface modification coating or overlay coating process wherein a substrate is immersed in a powder mixture, a so-called "pack," that consists of:

(1) The metallic powders that are to be deposited (usually aluminum, chromium, silicon, or combinations thereof);

(2) An activator (normally a halide salt); and

(3) An inert powder, most frequently alumina.

The substrate and powder mixture is contained within a retort that is heated to between 1030K to 1375K for sufficient time to deposit the coating.

(e) "Plasma spraying" is an overlay coating process wherein a gun (spray torch), which produces and controls a plasma, accepts powdered coating materials, melts them and propels them towards a substrate, whereon an integrally bonded coating is formed.

Note 1: "High velocity" means more than 750 meters per second.

Note 2: "Low pressure" means less than ambient atmospheric pressure.

(f) "Slurry deposition" is a surface modification coating or overlay coating process wherein a metallic or ceramic powder with an organic binder is suspended in a liquid and is applied to a substrate by either spraying, dipping or painting; subsequently, air or oven dried; and heat treated to obtain the desired coating.

(g) "Sputtering" is an overlay coating process wherein positively charged ions are accelerated by an electric field towards the surface of a target (coating material). The kinetic energy of the impacting ions is sufficient to cause target surface atoms to be released and deposited on the substrate.

Note: Triode, magnetron, or radio frequency sputtering to increase adhesion of coating and rate of deposition are ordinary modifications to the process.

(h) "Ion implantation" is a surface modification coating process in which the element to be alloyed is ionized, accelerated through a potential gradient and implanted into the surface region of the substrate. The definition includes processes in which the source of the ions is a plasma surrounding the substrate and processes in which ion implantation is performed simultaneously with "electron beam physical vapor deposition" or "sputtering."

(4) Design and production technical data, including software, for commodities that are listed below in numerical order by their respective Export Control Commodity Number. The commodities and related design and production technical data are controlled for nuclear weapons delivery reasons (see § 376.18).

ECCN 2018A: Specialized machinery, equipment, and gear for producing rocket systems (including ballistic missile systems, space launch vehicles, and sounding rockets) and unmanned air vehicle systems (including cruise missile systems, target drones, and reconnaissance drones) capable of delivering nuclear weapons (as defined in § 376.18(a)), their propulsion systems and components, and pyrolytic deposition and densification equipment.

ECCN 1118A: Production equipment for the development or production of rocket propellants.

ECCN 1131A: Pumps used in propulsion systems and related components as follows: pumps (except vacuum pumps), having all flow contact surfaces made of 90 percent or more tantalum, titanium or zirconium, either separately or combined, and when designed to operate in vibrating environments of more than 12g rms between 20 Hz and 2000 Hz, except when such surfaces are made of materials containing more than 97 percent and less than 99.7 percent titanium.

ECCN 1133A: Valves used in propulsion systems and related components as follows: servo valves designed for flow rates of 24 liters per minute or greater at a pressure of 250 bars, and having flow contact surfaces made of 90 percent or more tantalum, titanium or zirconium, either separately or combined, and when designed to operate in vibrating environments of more than 12g rms between 20 Hz and 2000 Hz, except when such surfaces are made of materials containing more than 97 percent and less than 99.7 percent titanium.

ECCN 1302A: Specially designed nozzles for producing pyrolitically derived materials formed on a mould, mandrel or other substrate from precursor gases that decompose in the 1,573K (1,300°C) to 3,173K (2,900°C) temperature range at pressures of 133.3 Pa to 19.995 kPa. (Commodities described in this entire entry.)

ECCN 1357A: Equipment for the production of fibers covered by ECCN 1763A, or their composites as follows, and specially designed components and accessories therefor. (Commodities described in this entire entry.)

ECCN 1361A: Commodities described in entire entry, except for paragraph (e) under the "List of Wind Tunnels Controlled by ECCN 1361A."

ECCN 1362A: Vibration test equipment. (Commodities described in this entire entry.)

ECCN 1385A: Specially designed production equipment for gyroscopes (gyros), accelerometers and inertial equipment con-trolled by ECCN 1485A. (Commodities described in this entire entry.)

ECCN 1460A: Commodities described in paragraphs (c), and (d) as applicable to (c), under the "List of Nonmilitary Equipment Controlled by ECCN 1460A" for lightweight turbojet and turbofan engines (including turbocompound engines) that are small and fuel efficient.

ECCN 1485A: Commodities described in this entire entry (Note: Department of State, Office of Munitions Control, has jurisdiction over certain inertial system technical data (see Interpretation 21 of Supplement No. 1 to § 399.2)).

ECCN 1501A: Commodities described in paragraphs (b)(2) through (5) and (c) under the "List of Navigation, Direction Finding, Radar and Airborne Communication Equipment Controlled by ECCN 1501A" for launch and ground support equipment, including precision tracking systems usable for complete rocket systems and unmanned air vehicle systems described in § 376.18(a).

ECCN 1516A: Commodities described in paragraph (c) under the "List of Receivers and Specialized Parts and Accessories Controlled by ECCN 1516A" for telemetering and telecontrol equipment usable for: 1. complete rocket systems and unmanned air vehicle systems described in § 376.18(a); and 2. launch and ground support of the above systems.

ECCN 1517A: Commodities described in paragraph (c) under the "List of Radio Transmitters and Components Controlled by ECCN 1517A" for telemetering and telecontrol equipment usable for: 1. complete rocket systems and unmanned air vehicle systems described in § 376.18(a); and 2. launch and ground support of the above systems.

ECCN 1518A: 1. Telemetering and telecontrol equipment usable for: 1. complete rocket systems and unmanned air vehicle systems described in § 376.18(a); and 2. launch and ground support of the above systems.

ECCN 1519A: Commodities described in paragraph (c) under the "List of Equipment Controlled by ECCN 1519A" for single- and multi-channel communications transmission equipment as follows: 1. telemetering and telecontrol equipment usable with complete rocket systems and unmanned air vehicle systems described in § 376.18(a), and 2. precision tracking systems for the above systems.

ECCN 1522A: Commodities described in paragraphs (b) and (c) under the "List of Lasers and Laser Systems Controlled by ECCN 1522A" as follows: 1. test and alignment equipment for flight control systems usable in complete rocket systems and unmanned air vehicle systems described in § 376.18(a); and 2. precision tracking systems for the above systems.

ECCN 1529A: 1. Commodities described in paragraph (a) under the "List of Equipment Controlled by ECCN 1529A" for launch and ground support equipment usable for complete rocket systems and unmanned air vehicle systems described in § 376.18(a); and 2. commodities described in paragraph (b)(4) under the List when part of a test system described in ECCN 1361A or 1362A.

ECCN 1531A: Commodities described in paragraphs (a) and (c) through (e) under the "Illustrative List of Frequency Synthesizers Controlled by ECCN 1531A" as follows: 1. avionics equipment usable in complete rocket systems and unmanned air vehicle systems described in § 376.18(a); 2. vibration test equipment (ECCN 1362A) and wind tunnels (ECCN 1361A); and 3. launch and ground support equipment usable for the systems described in § 376.18(a).

ECCN 1564A: A-D converters described in paragraph (d)(22)(a) under the "List of Electronic Component Assemblies, Sub-Assemblies, Printed Circuit Boards, and Microcircuits Controlled by ECCN 1564A" when usable in systems described in § 376.18(a) and having any of the following characteristics: rated for continuous operation at temperatures from below –45°C to above 55°C; designed to meet military specifications for ruggedized equipment, or modified for military use; or designed or radiation resistance.

ECCN 1565A: 1. Specially designed analog computers or specially designed hybrid (combined analog/digital) computers that are described in paragraphs (c) through (d), and (h) as applicable to (d), under the "List of Electronic Computers and Related Equipment Controlled by ECCN 1565A" for modeling, simulation, or design integration of the following: complete rocket systems and unmanned air vehicle systems described in § 376.18(a); and 2. technical data for analog or digital computers and related equipment that provide special capabilities as described in paragraphs (a), (b), and (f) under the "List of Electronic Computers and Related Equipment Controlled by ECCN 1565A."

ECCN 1568A: A-D converters described in paragraph (k) under the "List of Equipment Controlled by ECCN 1568A" when usable in systems described in § 376.18(a) and having any of the following characteristics: rated for continuous operation at temperatures from below –45°C to above 55°C; designed to meet military specifications for ruggedized equipment, or modified for military use; or designed for radiation resistance.

ECCN 1587A: Commodities described in paragraph (c) under the "List of Characteristics of Quartz Crystals and Assemblies Thereof Controlled by ECCN 1587A" when usable as launch and ground support equipment.

ECCN 1595A: Gravity meters (gravimeters), gravity gradiometers and specially designed components therefor, designed or modified for airborne or marine use, and having a static or operational accuracy of one milligal or better, with a time to steady-state registration of two minutes or less.

ECCN 1715A: Propellants and constituents as follows: high energy density fuels such as Boron Slurry, having an energy density of 40×10^6 joules/kg or greater.

ECCN 1801A: Propellants and constituents as follows: Polymeric substances, specifically: 1. carboxy-terminated polybutadiene (CTPB), and 2. hydroxy-terminated polybutadiene (HTPB).

PART 385

SPECIAL COUNTRY POLICIES AND PROVISIONS

§ 385.1
COUNTRY GROUP Z[1]: NORTH KOREA, VIETNAM, CAMBODIA AND CUBA

(a) As authorized by section 6 of the Export Administration Act of 1979 and by the Trading With the Enemy Act of 1917 as amended by Public Law 95–223, a validated license is required for foreign policy purposes for the export and reexport of virtually all U.S.-origin commodities and technical data to destinations in Country Group Z. Certain exceptions are contained in Parts 371 and 379, and in ECCN's 7599 and 7999 on the Commodity Control List. Except as noted below, the general policy is to deny all applications or requests to export or reexport U.S.-origin commodities and technical data to these destinations. Exports of donations to meet basic human needs may be authorized under a Humanitarian License, as described in § 373.5. Such exports may also be authorized for single transactions under an individual validated license. Exports to meet emergency needs that do not qualify for export under the Humanitarian License procedure will be considered on a case-by-case basis.

(b) In addition to the general policy set forth above, the following policies apply to Cuba.

(1) Cuba has been designated by the Secretary of State as a country that has repeatedly provided support for acts of international terrorism.

(2) On August 21, 1975, the U.S. Government announced modifications in those aspects of U.S. restrictions on trade with Cuba which affect third countries, in order to bring them into accord with the policy of the Organization of American States to allow each member state to determine for itself the nature of its economic and diplomatic relations with the Government of Cuba. In this context, the Department of Commerce generally will consider favorably on a case-by-case basis requests for authorization for the use of an insubstantial proportion of U.S.-origin materials, parts, or components in nonstrategic foreign-made products to be exported to Cuba, where local law requires, or policy in the third country favors, trade with Cuba. U.S.-origin content will generally be considered insubstantial when it amounts to 20 percent or less of the value of the product to be exported from the third country. Requests for authorization for the use of U.S.-origin parts, components or materials amounting to more than 20 percent by value of a foreign-made product to be exported to Cuba generally will not be approved. See § 376.12(e) for instructions on submission of requests for authorization.

§ 385.2
COUNTRY GROUP Q, W, and Y[1]: U.S.S.R., OTHER WARSAW PACT COUNTRIES, ALBANIA, MONGOLIAN PEOPLE'S REPUBLIC, AND LAOS

(a)(1) The Export Administration Act of 1979 states that it is the policy of the United States "to encourage trade with all countries with which we have diplomatic or trading relations, except those countries with which such trade has been determined by the President to be against the national interest." The Act also states that it is the policy of the United States "to restrict the export of goods and technology which would make a significant contribution to the military potential of any other country or combination of countries which would prove detrimental to the national security of the United States." Accordingly, and in compliance with the other sections of the Export Administration Act of 1979, the Department conducts a continuing review of commodities and technology to assure that prior approval is required for the export or reexport of U.S.-origin commodities and technical data to the U.S.S.R., Albania, Bulgaria, Czechoslovakia, Estonia, German Democratic Republic, Hungary, Laos, Latvia, Lithuania, Mongolian People's Republic, Poland, and Romania only if the commodities or technical data have a potential for being used in a manner that would prove detrimental to the national security of the United States. The general policy of the Department, however, is to approve applications or requests to export or reexport such commodities and technical data to these destinations (other than the U.S.S.R. and Poland) when the Department determines, on a case-by-case basis, that the commodities or technical data are for a civilian use or would otherwise not make a significant contribution to the military potential of the country of destination that would prove detrimental to the national security of the United States.

(2) To permit such policy judgments to be made, each export application and reexport request is reviewed in the light of prevailing policies with full consideration of all relevant aspects of the proposed transaction. The review generally includes an analysis of the kinds and quantities of commodities or technologies to be shipped; their military or civilian uses; the unrestricted availability abroad of the same or comparable items; the country of destination; the ultimate end-users in the country of destination; and the intended end-use.

(3) Applications covering certain commodities and technical data that are controlled by the United States

---

[1] See Supplement No. 1 to Part 370 for listing of Country Groups.

and certain other nations that cooperate in an international export control system and are proposed for export or reexport to Country Group Q, W, or Y may have to be forwarded to the Coordinating Committee (COCOM) of this international export control system for consideration in accordance with established COCOM procedures.

(4) Although each proposed transaction is considered individually, certain goods on the Commodity Control List are more likely to be approved than others. See the Advisory Notes for the applicable entry on the Commodity Control List (Supplement No. 1 to § 399.1) for an identification of such goods.

(b) An support of U.S. foreign policy to promote the observance of human rights throughout the world, a validated export license is required to export any commodity particularly useful in crime control and detection to any destination in Country Group Q, W, or Y.

(c) [Reserved]

(d) [Reserved]

(e) As authorized by section 6 of the Export Administration Act of 1979, a validated license is required for the export to the U.S.S.R. of technical data and equipment for the export to the U.S.S.R. of technical data and equipment for the manufacture of trucks, as defined in CCL entry 6398G, at the Kama River (Kam AZ) and ZIL truck plants. Licenses for such exports will generally be denied.

§ 385.3 [Reserved.]

§ 385.4

COUNTRY GROUPS T & V

(a) Republic of South Africa and Namibia

In conformity with the United Nations Security Council Resolutions of 1963 and 1977 relating to exports of arms and munitions to the Republic of South Africa, and consistent with U.S. foreign policy toward the Republic of South Africa and Namibia, the Department of Commerce has established the following special policies for commodities and technical data under its licensing jurisdiction.

(1) An embargo is in effect on the export or reexport to the Republic of South Africa and Namibia of arms, munitions, military equipment and materials, and materials and machinery for use in the manufacture and maintenance of such equipment. Commodities to which this embargo applies are listed in Supplement No. 2 to Part 379.

(2) An embargo is in effect on the export or reexport to the Republic of South Africa or Namibia of any commodity, including commodities that may be exported to any destination in Country Group V under a general license, where the exporter or reexporter knows or has reason to know that the commodity will be sold to or used by or for military or police entities in these destinations or used to service equipment owned, controlled or used by or for such military or police entities (See § 385.4(a)(7) and (10) for case-by-case exceptions.)

(3) An embargo is in effect on the export or reexport to the Republic of South Africa or Namibia of technical data—except technical data generally available to the public that meets the conditions of General License GTDA—where (i) the technical data relate to the commodities listed in Supplement No. 2 to Part 379 of this chapter, (ii) the exporter or reexporter knows or has reason to know that the technical data or any product of the data as defined in § 379.4(e) of this chapter are for delivery to or for use by or for the military or police entities of these destinations or for use in servicing equipment owned, controlled or used by or for these entities (see § 385.4(a)(7) and (10) for case-by-case exceptions), or (iii) the technical data consist of data to service computers or of computer software and the exporter or reexporter knows or has reason to know that it will be made available to or for use by or for apartheid enforcing entities identified in Supplement No. 1 to Part 385 (with case-by-case exceptions possible for humanitarian purposes). In addition, users in the Republic of South Africa or Namibia of technical data that do qualify for export or reexport under the provisions of General License GTDR must provide a written assurance to the U.S. exporter as required by § 379.4(e)(2) of this chapter.

(4) Parts, components, materials and other commodities exported from the United States under either a general or validated export license may not be used abroad to manufacture or produce foreign-made end-products where it is known or there is reason to know the end-product will be sold to or used by or for military or police entities in the Republic of South Africa or Namibia. (See § 385.4(a)(7) and (10) for case-by-case exceptions.)

(5) A validated export license is required for the export to the Republic of South Africa and Namibia of any instrument and equipment particularly useful in crime control and detection, as defined in § 376.14 of this chapter.

(6) General License GIT may not be used for any commodity destined for the Republic of South Africa or Namibia (see § 371.4(b) of this chapter).

(7) Applications for validated licenses will generally be considered favorably on a case-by-case basis for the export of medicines, medical supplies, medical equipment, related technical data, and parts and components, to any end-user.

(8) A validated license is required for export to all consignees of aircraft and helicopters. Applications will generally be considered favorably on a case-by-case basis for such exports for which adequate written assurances have been obtained against military, paramilitary, or police use.

(9)(i) A validated license is required for the export or reexport to government consignees of computers as defined in CCL entries 1565A and 6565G, and of goods to service or manufacture computers. "Goods to service or manufacture computers" include any national security-controlled item intended for such use, and any items in ECCN 6594F. Applications will be denied if the export is likely to be used by or for apartheid enforcing entities identified in Supplement No. 1 to Part 385, with case-by-case exceptions possible for humanitarian purposes. (ii) Insubstantial U.S. origin parts or peripherals in foreign origin computer systems may be approved on a case-by-case basis. Content is generally considered to be insubstantial when it is 20% or less by value. (iii) Applications to export computers or goods to service or manufacture computers to the Republic of South Africa or Namibia must be accompanied by the following certification signed by the ultimate consignee;

I (We) certify that we are the recipient of the commodities to be delivered under this license, that we are not affiliated with any apartheid-enforcing entity, and that the commodities will not be sold or otherwise made available, directly or indirectly, to or for use by or for the following entities in the Republic of South Africa and Namibia: police or military entities, any entity involved directly or indirectly in either a nuclear or sensitive nuclear end use, or entities identified by the U.S. Department of State as enforcing apartheid as reflected in Supplement No. 1 to Part 385 of the Export Administration Regulations. These commodities are not to be used to service computers owned, controlled, or used by or for the entities indicated above, or used to manufacture computers intended for such entities. I (We) will cooperate with post-shipment inquiries by U.S. officials to verify disposition or use of the commodities. If requested by the exporter, we will periodically provide information concerning the disposition or use of commodities received under this license, including the identity of customers to whom the items were resold.

Note.—See §§ 373.3(a)(2) and 378.3 for definitions relating to nuclear end uses.

(iv)(A) Exporters shall make available to each ultimate consignee a current copy of Supplement No. 1 to Part 385 identifying apartheid enforcing entities. OEL may require the licensee to certify subsequent to delivery that the computers have not been retransferred to unauthorized end-uses or end-users.

(B) When the license authorizes resale within the Republic of South Africa or Namibia, the requirement may include identification of customers to whom the ultimate consignees have sold the computers, with further identifications of customers to be provided at six month intervals until all computers exported under the license have been sold.

(10) Applications for validated licenses will generally be considered favorably on a case-by-case basis for the export of commodities and related technical data, and parts and components, to be used in efforts to prevent acts of unlawful interference with international civil aviation.

(11) Applications for validated licenses for commodities and technical data for nuclear production or utilization facilities or those likely to be diverted to such facilities will be denied, except on a case-by-case basis to assist International Atomic Energy Agency (IAEA) safeguards or IAEA programs generally available to member states, or for technical programs to reduce proliferation risks, or for exports that are determined to be necessary for humanitarian reasons to protect the public health and safety.

(12) License applications involving contracts entered into prior to the President's Executive Order of September 9, 1985, will be considered on a case-by-case basis in accordance with the regulations and policies that were in effect prior to the issuance of such Order. Applications involving contracts entered into on or after September 9, 1985, will be subject to the October 11, 1985, revisions to the regulations that implement that Order. Applications involving contracts for equipment and technical data intended to manufacture computers entered into on or after October 2, 1986 will be subject to the restrictions imposed pursuant to the Comprehensive Anti-Apartheid Act of 1986. Those involving contracts entered into prior to that date

will be considered on a case-by-case basis consistent with the purposes of that Act.

(13) Pursuant to Section 321 of the Comprehensive Anti-Apartheid Act of 1986, a validated license is required for the export from the United States to the Republic of South Africa and Namibia of crude oil (ECCN 4781B) and refined petroleum products (ECCNs 4782B, 4783B, and 4784B). License applications for these commodities will be denied. In addition, reexport authorization is required for any export of such commodities from outside the United States to these destinations if the commodities originated in the United States and are being exported by a person or firm subject to the jurisdiction of the United States. For purposes of this paragraph, the term "person subject to the jurisdiction of the United States" means:

(i) Any U.S. citizen or permanent resident alien, except if acting in the course of employment by a juridical person organized under the laws of a foreign jurisdiction;

(ii) A juridical person organized under the laws of the United States; or

(iii) Any person in the United States, defined to include those territories listed in § 370.2.

The reexport provisions of Part 374 and the provisions of § 376.12 are not applicable to the controls covered by this § 385.4(a)(13), except in the case of reexports to South Africa and Namibia by a person subject to the jurisdiction of the United States as defined above. However, the export of these commodities from the United States to any destination with knowledge that they will be reexported, directly or indirectly, in whole or in part, to the Republic of South Africa and Namibia is prohibited. Pursuant to section 604 of the CAA, no person may undertake or cause to be undertaken any transaction or activity with the intent to evade the restrictions described herein.

(c) People's Republic of China

(1) The general licensing policy is to consider exports for China under the Country Groups T and V policies set forth in paragraph (g) below, except that there are certain commodities, data, and end-uses that may require extended review or denial. Of particular concern are exports that would make a direct and significant contribution to nuclear weapons and their delivery systems, electronic and anti-submarine warfare, intelligence gathering, power projection, and air superiority. Licenses may be approved even when the end-user or end-use is military. Commodities or data may be approved for

export even though they may contribute to Chinese military development.

(2) Each application will be considered individually. The Advisory Notes in the CCL (Supplement No. 1 to § 399.1) headed "Note for the People's Republic of China" provide guidance on equipment likely to be approved most rapidly for China and that for the most part will not require any interagency review. Items with higher performance levels than those described in the Advisory Notes may also be approved on a case-by-case basis.

(3) Applications covering commodities and technical data approved by the United States that are controlled by the COCOM international export control system may have to be forwarded to COCOM for consideration in accordance with established procedures before a license is issued.

(d) People's Democratic Republic of Yemen,
Syria, and Iran

(1) A validated license is required for foreign policy purposes for the export to the People's Democratic Republic of Yemen, Syria, or Iran (countries that have repeatedly provided support for acts of international terrorism) of crime control and detection equipment (see § 376.14), military vehicles and items specially designed to produce military equipment as defined in CCL entries 2018A, 1118A, 2406A, and 2603A (see § 376.16), and certain other commodities as specified below.

(2) For the People's Democratic Republic of Yemen, a validated license is required for the export of aircraft valued at $3 million or more and helicopters as defined in CCL entries 1460A(a), 1460A(b), 2460A and 5460F, except aircraft and helicopters for use by regularly scheduled airlines based in the People's Democratic Republic of Yemen for which assurances have been submitted to OEL against military use, and of goods or technology subject to national security controls if the export is destined to military end-users or for military end-use and is valued at $7 million or more. In the case of the use abroad of U.S.-origin parts, components or materials in foreign-origin products, the dollar limits set forth above apply to the U.S. content. Licenses and authorizations for helicopters 10,000 pounds empty weight or less will be approved when the export is in performance of a contract or agreement entered into before April 28, 1986, but subject to applicable national security or other applicable controls. Applications for validated licenses will be considered on a case-by-case basis to determine whether issuance of a license would be consistent with the provisions of section 6 and the

applicable policies set forth in section 3 of the Act (exports subject to national security controls must also meet the national security provisions of the Act). When the request for authorization involves use of U.S.-origin parts, components, or materials in foreign-origin products destined for the People's Democratic Republic of Yemen, licensing decisions will take into account whether the U.S. content is 20% or less by value.

(3)(i) For Iran, a validated license is required for the export of all aircraft and helicopters, and related parts and components, and of marine outboard engines with a horsepower of 45 or more, as defined in CCL entries 1460A, 2460A, 4460B, 5460F, 6460F, 1485A, 6494F, and 1501A(a), (b)(1) and (c)(1); and of all goods and technical data subject to national security controls if the export is destined to a military end-user or for military end-use. Applications for export to Iran of commodities and technology subject to these controls will generally be denied. However, applications may be considered on a case-by-case basis, if:

(A) The transaction involves the export or reexport of goods or technical data under a contract that was in effect before:

(1) January 23, 1984, in the case of helicopters over 10,000 lbs. empty-weight, aircraft valued at $3 million or more each, or national security controlled items valued at $7 million or more; or

(2) September 28, 1984, in the case of all other commodities or technical data.

(B) The commodities or technical data had been exported from the United States before January 23, 1984 or September 28, 1984 as appropriate.

(C) The U.S. content of foreign-produced commodities is 20% or less by value.

(ii) Applicants who wish such factors to be considered in reviewing their license applications must submit adequate documentation demonstrating the value of the U.S. content, the existence of the pre-existing contract, or the date of export from the U.S.

(4)(i) For Syria, a validated license is required for the export of all aircraft and helicopters, and related parts and components, as defined by CCL entries 1460A, 2460A, 4460B, 5460F, 6460F, 1485A and 1501A(a), (b)(1) and (c)(1), and all goods and technical data subject to national security controls. For purposes of this § 385.4(d)(4), technical data subject to national security controls are those data set forth in § 379.4(c) and (d). Applications for exports to Syria of commodities and technical data

subject to these controls will generally be denied. However, applications will be considered favorably on a case-by-case basis if:

(A) The transaction involves the export or reexport of goods or technical data under a contract involving shipment to Syria that was in effect before:

(1) April 28, 1986, in the case of helicopters 10,000 pounds empty weight or less; or

(2) December 16, 1986, for all other commodities except those described in paragraph (d)(4)(iii) of this section; or

(B) The transaction involves the reexport to Syria of goods where Syria was not the intended ultimate destination at the time of the original export from the United States, if—

(1) In the case of helicopters 10,000 pounds empty weight or less, they had been exported from the U.S. prior to April 28, 1986; or

(2) In the case of other commodities, except those described in paragraph (d)(4)(iii) of this section, they had been exported from the U.S. before June 18, 1987; or

(C) The U.S. content of foreign-produced commodities is 20% or less by value; or

(D) The commodities are medical equipment.

(ii) Applicants who wish such factors to be considered in reviewing their license applications must submit adequate documentation demonstrating the value of the U.S. content, the existence of the pre-existing contract, the specifications and intended medical use of the equipment, or the date of export from the United States.

(iii) The favorable consideration policies identified in section 6(j) of the Act will be notified set forth in § 385.4(d)(4) do not apply to aircraft valued at $3 million or more or to helicopters exceeding 10,000 pounds empty weight (unless they are destined for a regularly scheduled airline with assurance against military use), or to national security controlled goods valued at $7 million or more destined for military end-users or end-uses.

(5) The appropriate Congressional Committees identified in section 6(j) of the Act will be notified 30 days before any application falling under this subsection valued at $1 million or more is approved.

(e) Iran, Iraq and Syria

In support of U.S. foreign policy, and particularly U.S. policies of opposing prohibited use of chemical weapons and maintaining neutrality in the Iran/

Iraq war and of promoting a mediated end to that war, an individual validated license is required to export from the United States N,N-diisopropylaminoethane-2-thiol, N,N-diisopropylaminoethyl-2-chloride, dimethylamine, dimethylamine hydrochloride, dimethyl phosphite (dimethyl hydrogen phosphite), ethylene chlorohydrin (chloroethanol), 3-hydroxy-1-methylpiperidine, phosphorous trichloride, potassium fluoride, 3-quinuclidinol, thionyl chloride, and trimethyl phosphite to Iran, Iraq, and Syria. Applications for validated licenses will be considered on a case-by-case basis. Applications will generally be denied where there is reason to believe that these chemicals will be used in producing chemical weapons or will otherwise be devoted to chemical warfare purposes. However, applications for export of dimethylamine, dimethylamine hydrochloride, ethylene chlorohydrin (chloroethanol), and potassium fluoride to Syria will generally be approved when the export is in performance of a contract or agreement entered into before April 28, 1986. Applications for export of N,N-diisopropylaminoethane-2-thiol; N,N-diisopropylaminoethyl-2-chloride; dimethyl phosphite (dimethyl hydrogen phosphite); 3-hydroxy-1-methylpiperidine; phosphorous trichloride; 3-quinuclidinol; thionyl chloride; and trimethyl phosphite to Iran, Iraq, and Syria will generally be approved when the export is in performance of a contract or agreement entered in before July 6, 1987. In the absence of a contract as described above, applications to export chemicals listed in this section to Iran, Iraq, or Syria will generally be denied. The reexport provisions of Part 374 and the provisions of § 376.12 are not applicable to the foreign policy controls covered by paragraph (e) of this section. However, the export of these commodities from the United States to any destination with knowledge that they will be reexported, directly or indirectly, in whole or in part, to Iran, Iraq or Syria is prohibited without a validated license.

(f) Afghanistan

The Soviet military presence in Afghanistan requires special attention to exports because of the likelihood that commodities or technical data entering Afghanistan will be available to the U.S.S.R. Accordingly, the validated licensing requirements for the U.S.S.R. extend to shipments to Afghanistan. The purpose of this action is not to deny commodities or technical data to Afghanistan but to implement more effectively those national security and foreign policy controls already in existence with regard to the U.S.S.R. Accordingly, the statutory bases for controlling such shipments to Af-

ghanistan parallel those for the U.S.S.R. With regard to applications for shipments subject to national security controls, the general policy is to deny such applications if they would be denied if destined for the U.S.S.R.

(g) Other Countries in Group T and V

(1) General Policy. For other countries in Country Group T and V, the Department of Commerce requires prior approval for the export or reexport of selected commodities and technical data that include—

(i) Commodities the United States and other Free World governments have agreed to control vis a vis their export to Communist destinations in Europe and the Far East, and

(ii) Certain other commodities that the United States considers to have high potential for strategic use, including nuclear-related commodities. Technical data relating to a few commodities of particular strategic significance also are subject to the prior approval procedure. (See § 379.4 for technical data controls, which may vary with country of destination and type of data.)

(iii) The policy generally is one of approval of applications unless there is a significant risk that the commodities or data will be used or diverted contrary to the objectives of a specific U.S. export control program.

(2) Crime control and detection commodities. In support of U.S. foreign policy to promote the observance of human rights throughout the world, an individual validated export license is required to export any crime control and detection equipment as defined in § 376.14 to any destination in these country groups except Australia, Belgium, Denmark, France, the Federal Republic of Germany (including West Berlin), Greece, Iceland, Italy, Japan, Luxembourg, the Netherlands, New Zealand, Norway, Portugal, Turkey and the United Kingdom.

(3) Regional stability commodities and equipment. In support of U.S. foreign policy to maintain regional stability, an individual validated export license is required to export military vehicles and certain commodities used to manufacture military equipment identified in § 376.16, to any destination in these country groups except Australia, Belgium, Denmark, France, the Federal Republic of Germany (including West Berlin), Greece, Iceland, Italy, Japan, Luxembourg, the Netherlands, New Zealand, Norway, Portugal, Turkey and the United

Kingdom. (See § 385.7 for regional stability controls on exports to Libya.)

§ 385.5

[Reserved]

§ 385.6

CANADA

Except as indicated below, the general policy is to permit shipments of commodities and technical data to Canada for consumption or use in that country without an export license. When the commodities or technical data are transiting Canada or are intended for reexport from Canada to another foreign destination and such shipment would require a validated license if made directly from the United States to that destination, an export license or reexport authorization is required. The licensing action will be based on the policy applicable to a direct shipment from the United States to such other destination. (See §§ 374.1 and 386.1(d) for commodities in transit via Canada.) A validated license also is required for export to Canada if—

(a) The technical data are described in § 379.4(c) or § 379.5(e) unless the technical data may be exported under the provisions of General License *GTDA;*

(b) The commodity is related to nuclear weapons, nuclear explosive devices, nuclear testing, the chemical processing of irradiated special nuclear or source material, the production of heavy water, the separation of isotopes of source and special nuclear material, or the fabrication of nuclear reactor fuel containing plutonium, as described in § 378.3; or

(c) The Commodity Control List (Supplement No. 1 to § 399.1) indicates that a validated license is required for export to Canada.

§ 385.7

COUNTRY GROUP S: LIBYA

As authorized by Section 6 of the Export Administration Act of 1979, the following special policies and procedures for commodities and technical data are in effect for Libya for foreign policy purposes.

Note:

The Libyan Sanctions Regulations (31 CFR Part 550) administered by the Department of the Treasury restrict exports from the United States to Libya. As reflected in the General Order contained in Section 390.7, effective February 1, 1986, a license issued under the Treasury Regulations will constitute an authorization under the Export Administration Regulations for an export from the United States. No license application need be filed with Commerce. Shipments to Libya from foreign countries that are subject to the provisions of Part 374, §§ 376.12 and 379.8 and this section and are not subject to the Libyan Sanctions Regulations continue to require authorization under the Export Administration Regulations.

(a) Except as stated below, a validated license or reexport authorization is required for all U.S.-origin commodities or technical data, as well as foreign produced products of U.S. technical data exported from the United States after March 12, 1982 subject to national security controls for which written assurances against shipments to Libya are required under § 379.4 of the Export Administration Regulations. Excepted from this licensing requirement are those goods exported under the Humanitarian License procedure outlined in § 373.5 and commodities and data exported pursuant to special general licenses described in Parts 371 and 379. License applications and reexport requests will be reviewed in accordance with the licensing policies stated below.

(1) Licenses will generally be denied for—

(i) Items controlled for national security purposes and related technical data, including controlled foreign produced products of U.S. technical data exported from the United States after March 12, 1982; and

(ii) Oil and gas equipment and technical data, if determined by the Office of Export Licensing not to be readily available from sources outside the United States; and

(iii) Goods and technical data destined for the petrochemical processing complex at Ras Lanuf, where such items would contribute directly to the development or construction of that complex. Items destined for the township at Ras Lanuf, or for the public utilities or harbor facilities associated with that township, generally will *not* be regarded as making such a contribution where their functions will be primarily related to the township, utilities or harbor.

(2) Notwithstanding the presumptions of denial in paragraph (1) above:

(i) Licenses and authorizations generally will be issued when the transaction involves—

(A) The export or reexport of commodities or technical data under a contract in effect prior to March 12, 1982, where failure to obtain a license would not excuse performance under the contract;

(B) Reexport of goods or technical data not controlled for national security purposes that had been exported from the United States prior to March 12, 1982, or exports of foreign products incorporating such items as components; or

(C) Use of U.S. origin parts, components, or materials in foreign origin products destined for Libya, where the U.S. content is 20 percent or less by value.

(ii) Licenses and authorizations will generally be considered favorably on a case-by-case basis when the transaction involves—

(A) Reexports of goods or technical data subject to national security controls that were exported prior to March 12, 1982 and exports of foreign products incorporating such U.S.-origin components, where the particular authorization would not be contrary to specific foreign policy objectives of the United States; or

(B) Items destined for use in the development or construction of the petrochemical processing complex at Ras Lanuf, where the transaction could be approved but for the general policy of denial set out in paragraph (a)(1)(iii) above, and where either—

(1) The transaction involves a contract in effect before December 20, 1983 that requires export or reexport of the goods or technical data in question; or

(2) The goods or technical data had been exported from the U.S. before that date.

(C) Other unusual situations such as transactions involving firms with contractual commitments in effect before March 12, 1982.

(3) All other exports and reexports will generally be approved, subject to any other licensing policies applicable to a particular transaction.

(b) Libya has been designated by the Secretary of State as a nation repeatedly supporting acts of international terrorism. For licensing requirements relating to Anti-Terrorism Controls see § 385.4(d).

(c) A validated license is required for the export to Libya of off-highway wheel tractors of carriage capacity of 10 tons or more, as defined in CCL entry 6490F. Applications for validated licenses will generally be considered favorably on a reasonable case-by-case basis for the export of such tractors in reasonable quantities if for civil use.

(d) A validated license is required for foreign policy purposes for the export or re-export to Libya of any aircraft (including helicopters) and any parts and accessories controlled under ECCNs 1460A, 2460A, 4460B, 5460F, 6460F, 1485A, and 1501A(a), (b)(1) and (c)(1). This control includes any such aircraft parts intended for use in the manufacture, overhaul, or rehabilitation in any third country of aircraft that will be exported or reexported to Libya or Libyan nationals. Applications for validated licenses will generally be approved on a case-by-case basis for aircraft unlikely to be diverted to military use because they are destined to a priority civil use. Applications will generally be denied for exports that would constitute a high risk of increasing Libyan capabilities to carry military cargo or troops or to conduct military reconnaissance or observation missions.

(e) In appropriate cases, applications for licenses and requests for extensions of licenses under paragraph (a) above must be accompanied by a certified true copy of a contract entered into prior to March 12, 1982. Where the custom in an industry is to rely upon agreements other than contracts, substitute documentation will be considered on a case-by-case basis. Requests for authorization to reexport must be accompanied by shipping documents establishing the date of export from the United States.

(f) Commodities and technical data subject to more than one type of control (e.g., national security, anti-terrorism, regional stability, nuclear nonproliferation) will be reviewed under all applicable standards. The most restrictive standard will be applied.

SOUTH AFRICAN ENTITIES ENFORCING APARTHEID

The following have been identified as South African entities that enforce apartheid. Exporters should be aware that this list cannot be all-inclusive because names of agencies are subject to change, and because agencies may assume apartheid enforcing activities in the future. Before making commitments to export, exporters may wish to seek guidance from the Office of Southern African Affairs (AF/S), Department of State, Washington, DC 20520, to ascertain whether or not a potential customer is an "agency enforcing apartheid."

Ministry of Justice

Ministry of Home Affairs and National Education

Ministry of Constitutional Development and Planning

Ministry of Law and Order

Ministry of Manpower

Department of Public Works and Land Affairs

Ministry of Education and Developing Aid, including the Development Boards and the Rural Development Boards (formerly known as the Ministry of Cooperation, Development and Education) but excluding the Department of National Education and Training

Other agencies enforcing apartheid including local, regional and "Homeland agencies" e.g., those that regulate employment, classification, or residence of non-whites.

Note.—The Department of State has determined that the following agencies are not considered to be apartheid enforcing entities:

Ministry of Communication and Public Works (This includes post and telecommunication agencies)

Ministry of Agricultural Economics and Water Affairs

Ministry of Mineral and Energy Affairs

Ministry of Finance.

Department of Transportation

Special Country Policies and Provisions   Supplement No. 2 to Part 385

INTERPRETATIONS

1. The Department has received inquiries as to whether certain entities in the Republic of South Africa and Namibia are considered police or military entities and hence subject to the embargo policy set forth in § 385.4.

(a) In addition to the military and police of South Africa and Namibia, the following are considered to be police or military entities:

ARMSCOR (Armaments Development and Production Corporation) and its subsidiaries: Nimrod, Atlas Aircraft Corporation, Eloptro (Pty) Ltd., Kentron (Pty) Ltd., Infoplan Ltd., Lyttleton Engineering Works (Pty) Ltd., Naschem (Pty) Ltd., Pretoria Metal Pressings (P.M.P.) (Pty) Ltd., Somchem (Pty) Ltd., Swartklip Products (Pty) Ltd., Telecast (Pty) Ltd., and Musgrave Manufacturers and Distributors,

Department of Prisons

"Homeland" Police and Armed Forces

National Institute for Aeronautics & Systems Technology (NIAST) of the Council for Scientific and Industrial Research (CSIR)

National Intelligence Services

South African Railways Police Force.

Weapons Research activities of the Council for Scientific and Industrial Research (CSIR)

(b) It is also the Department's position that certain municipal and provincial law enforcement officials, such as traffic inspectors and highway patrolmen, although separate from the South African police, are police entities because they have functions that are performed by the police in the United States. Other law enforcement entities and officials that do not have functions performed by the police in the United States, such as the Department of Customs, Department of Justice, health inspectors and licensing authorities, are not considered police entities.

Form ITA-622P

APPLICATION FOR EXPORT LICENSE

(See following pages)

FORM ITA-622P, APPLICATION FOR EXPORT LICENSE

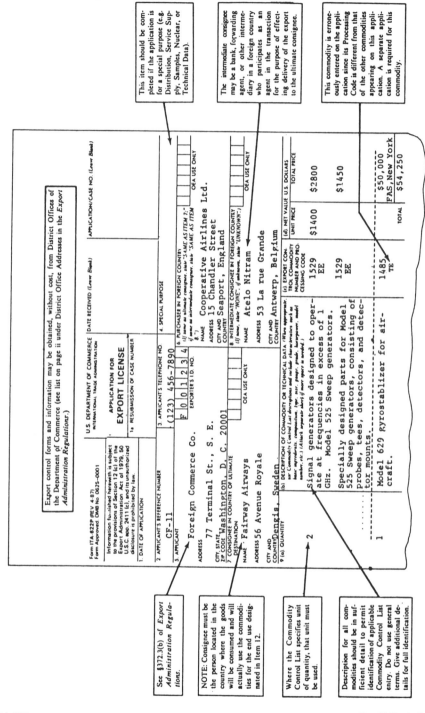

This item should be completed if the application is for a special purpose (e.g. Distribution, Service Supply, Samples, Nuclear, or Technical Data).

The intermediate consignee may be a bank, forwarding agent, or other intermediary in a foreign country who participates as an agent in the transaction for the purpose of effecting delivery of the export to the ultimate consignee.

This commodity is erroneously entered on the application since its Processing Code is different from that of the other commodities appearing on this application. A separate application is required for this commodity.

See §372.3(b) of *Export Administration Regulations.*

NOTE: Consignee must be the person located in the country where the goods will be consumed and will actually use the commodities for the end use designated in Item 12.

Where the Commodity Control List specifies unit of quantity, that unit must be used.

Description for all commodities should be in sufficient detail to permit identification of applicable Commodity Control List entry. Do not use general terms. Give additional details for full identification.

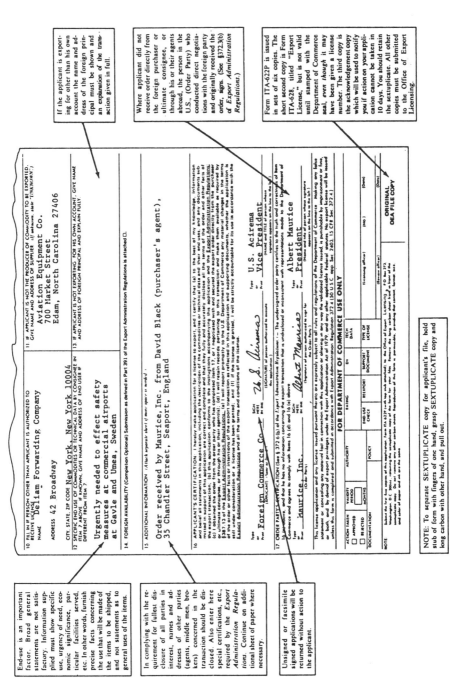

If the applicant is exporting for other than his own account the name and address of the foreign principal must be shown and an explanation of the transaction given in full.

Where applicant did not receive order directly from the foreign purchaser or ultimate consignee, or through his or their agents abroad, the person in the U.S. (Order Party) who conducted direct negotiations with the foreign party and originally received the order, signs. (See §372.3(b) of *Export Administration Regulations*.)

Form ITA-622P is issued in sets of six copies. The short second copy is Form ITA-628, titled "Export License," but is not valid until stamped with the Department of Commerce seal, *even though* it may have been given a license number. The third copy is the acknowledgement copy which will be used to notify you if action on your application cannot be taken in 10 days. You should retain the sextuplicate. All other copies must be submitted to the Office of Export Licensing.

End-use is an important factor. Broad general statements are not satisfactory. Information supplied must show specific use, urgency of need, economic significance, particular facilities served, etc. In other words, furnish precise facts concerning the use that will be made of the items to be shipped, and not statements as to general uses of the items.

In complying with the requirement for fullest disclosure of all parties in interest, names and addresses of other parties (agents, middle men, brokers) concerned in the transaction should be disclosed. Also enter here special certifications, etc., required by the *Export Administration Regulations*. Continue on additional sheet of paper where necessary.

Unsigned or facsimile signed applications will be returned without action to the applicant.

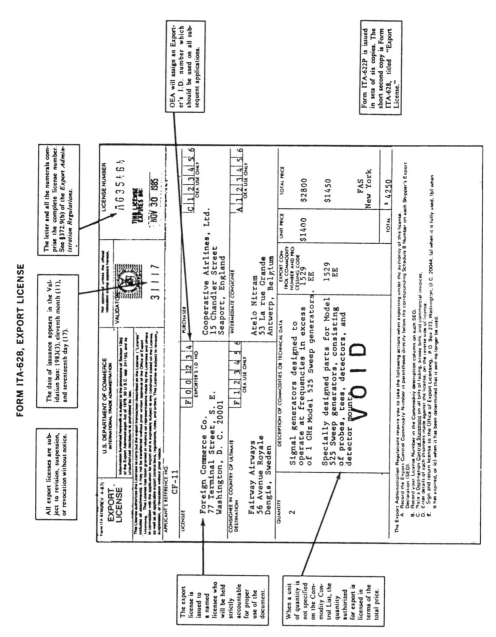

FORM ITA-628, EXPORT LICENSE

All export licenses are subject to revision, suspension, or revocation without notice.

The date of issuance appears in the Validation box: 1983(3), eleventh month (11), and seventeenth day (17).

The letter and all the numerals comprise the complete license number. See §372.9(b) of the *Export Administration Regulations.*

OEA will assign an Exporter's I.D. number which should be used on all subsequent applications.

Form ITA-622P is issued in sets of six copies. The short second copy is Form ITA-628, titled "Export License."

The export license is issued to a named licensee who will be held strictly accountable for proper use of the document.

When a unit of quantity is not specified on the Commodity Control List, the quantity authorized for export is licensed in terms of the total price.

U.S. DEPARTMENT OF COMMERCE
INTERNATIONAL TRADE ADMINISTRATION

EXPORT LICENSE

LICENSE NUMBER: A 6 3 5 4 6 4

THIS LICENSE EXPIRES ON: NOV 30 1985

VALIDATOR: 3 1 1 1 7

APPLICANT'S REFERENCE NO. CF-11

LICENSEE
Foreign Commerce Co.
77 Terminal Street, S. E.
Washington, D. C. 20001

EXPORTER'S I.D. NO. F 0 0 1 2 3 4

CONSIGNEE IN COUNTRY OF ULTIMATE DESTINATION
Fairway Airways
56 Avenue Royale
Dengis, Sweden

OEA USE ONLY F 1 2 3 4 5 6

PURCHASER
Cooperative Airlines, Ltd.
15 Chandler Street
Seaport, England

OEA USE ONLY C 1 2 3 4 5 6

INTERMEDIATE CONSIGNEE
Atelo Nitram
53 La rue Grande
Antwerp, Belgium

OEA USE ONLY A 1 2 3 4 5 6

QUANTITY	DESCRIPTION OF COMMODITIES OR TECHNICAL DATA	EXPORT CONTROL COMMODITY NUMBER AND PROCESSING CODE	UNIT PRICE	TOTAL PRICE
2	Signal generators designed to operate at frequencies in excess of 1 GHz Model 525 Sweep generators.	1529 EE	$1400	$2800
	Specially designed parts for Model 525 Sweep generators, consisting of probes, tees, detectors, and detector mount	1529 EE		$1450
				FAS New York
			TOTAL	$ 4250

VOID

The Export Administration Regulations require you to take the following actions when exporting under the authority of this license.
A. Record the Export Control Commodity Number in parentheses directly below the corresponding Schedule B Number on each Shipper's Export Declaration (SED).
B. Record your License Number in the Commodity description column on such SED.
C. Place a Destination Control Statement on all bills of lading, air way bills, and commercial invoices.
D. Enter details of each shipment made against the license, on the reverse side of license.
E. Sign and return license to the Office of Export Licensing, P.O. Box 273, Washington, D. C. 20044, (a) when it is fully used, (b) when it has expired, or (c) when it has been determined that it will no longer be used.

REVERSE OF FORM ITA-628P, EXPORT LICENSE

RECORD OF SHIPMENTS

Each shipment made against this Export License shall be recorded below. If more space is needed use a continuation sheet. (See Export Administration Regulations §386.2(d).) Shipping tolerances apply only to the unshipped balance remaining on the license at the time of shipment (§386.7).

QUANTITY SHIPPED	DESCRIPTION OF COMMODITIES	DOLLAR VALUE	NAME OF EXPORTING CARRIER	POINT OF EXIT OR POST OFFICE OF MAILING	DATE OF EXPORT	INITIALS OF PERSON MAKING ENTRY
3*	Signal Generators Model 525	4200*	ss Selfont	N.Y., N.Y.	12/1/84	*L.M.*
	Pts. for Model 525 Sweep Generator	1450	ss Selfont	N.Y., N.Y.	12/1/84	*L.M.*

I certify that shipments have been made under this license as indicated above. There is no material or substantive change in facts on which the license or subsequent amendments were issued.

Richard Mann
Signature of Licensee
or duly authorized agent

Delium Forwarding Co.
42 Broadway, N.Y., New York 10004
Address

12/1/84
Date

*Note that the quantity and value originally licensed was amended to permit this export. See page 19 of the Forms Supplement.

Section 3

Technology Search—
A Screening Approach

As we discussed in the section on Technology Acquisition, there are many approaches to searching for a desired technology. In Chapter 12, we considered general categories of searches, such as Defined Product, Defined Area, and General Area technology searches. Under these categories, we discussed various approaches. There are, for example, a number of capable firms and consultants who specialize in technology searches and evaluations. In addition, others prefer to make use of personal and professional contacts in undertaking a technology search. Still others will use a combination of these and other approaches that were discussed in Chapter 12.

With most of these technology search approaches, however, a preliminary screening can be very useful in providing a first-pass perspective of "what's out there" in the published literature. One example of this particular screening technique is the use of published searches available from the National Technical Information Service (NTIS).

NTIS, an agency of the U.S. Department of Commerce, manages an information collection of approximately two million titles, a considerable number of which contain foreign technology or marketing information. In addition, NTIS provides an annual catalog of published searches in many different fields, which are available at a very reasonable cost from this organization. These not only access the comprehensive NTIS database of federally sponsored research, but also cover twenty other large technologically oriented databases. The following three pages outline the databases covered by these published searches.

Published Searches Give You a Head Start When You Need Scientific and Technical Information and Analyses

NTIS Offers You Searches of Numerous Important Databases

To multiply the potential benefits available to you through Published Searches, NTIS offers subject searches from other databases. Most databases now include abstracts as part of the bibliographic record but, in the beginning, a few did not. Therefore, some searches may contain titles only for items indexed during the early years of a file. World Textile Abstracts (WTA) did not add abstracts before January, 1983.

NTIS Bibliographic Database — As the central information source for federally sponsored research, NTIS receives some 250 research reports daily. Major corporations, trade associations, and university and private research facilities all contribute their results to NTIS. Some 350 Government sources include: NASA, National Bureau of Standards, Department of Energy, and Department of Defense. The NTIS Bibliographic Database also contains references to Government-generated software packages, statistical data files (available on magnetic tape as well as floppy diskette) and other NTIS subscription products. The database carries over one million records. 70,000 reports are added each year. The file is updated twice monthly and dates back to 1964.

American Petroleum Institute (API) — Covers worldwide journal and preprint literature since 1964 on topics such as petroleum processing, fuels and lubricants, petrochemicals, petroleum transportation and storage, environmental matters, and non-petroleum energy sources.

Engineering Information Inc. (EI) — Covers international journals, articles, and reports on prominent research and applications of technical ideas relevant to the field of engineering.

Food Science and Technology Abstracts (FSTA) — FSTA is the main database of International Food Information Service (IFIS). It covers the international literature since 1969 on all matters concerned with research and development in food science and technology.

U.S. Patent Bibliographic Database — The U.S. Patent and Trademark Office (PTO) produces and leases many machine-readable patent databases that are processed and offered to the public through commercial systems. The primary patent database, the Patent Full Text Database, has several subfiles. One of the subfiles, the Patent Bibliographic Database with Exemplary Claims, is used to produce Published Searches. *They are not to be* construed as "legal" patent searches.

International Aerospace Abstracts (IAA) — Contains more than 850,000 citations to worldwide articles related to aerospace research and technology and allied sciences.

Life Sciences Collection (LSC) — Contains 770,000 records updated monthly, with approximately 8,000 new entries covering microbiology, biochemistry, genetics, ecology, toxicology, entomology, animal behavior, immunology, neurosciences, and chemoreception.

Energy Data Base (EDB)	Contains more than 1 million references to worldwide scientific and technical information of interest to the U.S. Department of Energy; reports, journals, patents, translations, and conference proceedings. Covers all aspects of international research on nuclear science, coal, and biomass.	**RAPRA (The Database of Rubber and Plastics Research Association of Great Britain)**	Covers journal articles (approximately 60 percent of database), patents, trade literature, Government reports, etc., on chemistry and the chemical engineering of polymers as well as the technical and commercial aspects of rubber and plastics. The database contains approximately 240,000 items, with 23,000 added annually.
Computer Database	Contains comprehensive information pertaining to the computer, telecommunication and electronic industries. The file dates back to 1983 and includes over 200,000 records. Source material drawn from journals, newsletters, and conference proceedings with 2500 new records added monthly.	**Oceanic Abstracts (OCEANIC)**	Covers the worldwide literature in the field of marine research, with topics including marine geology, biology, geophysics, pollution, and marine resources (living and nonliving). Each year approximately 10,500 items are added to the database, which at present consists of approximately 165,000 citations.
Conference Papers Index (CONF)	Contains more than 1,040,000 records with 3,000 added monthly. Covers the literature presented annually at regional, national, and international meetings on the topics of engineering, and physical sciences.	**Pollution Abstracts (POLLUTION)**	Contains 119,000 entries dealing with worldwide coverage of pollution-related topics such as air and water pollution, solid wastes, noise, pesticides, radiation, and general environmental quality.
Management Contents	Contains over 230,000 citations with almost 2,000 new documents being added each month. Over 150 business and law journals, newsletters, tabloids, proceedings and transactions are indexed and abstracted cover-to-cover.	**Research Association for the Paper & Board, Printing & Packaging Industries (PIRA)**	Database of approximately 110,000 items, covering mainly journal articles from industry-specific scientific, technical, marketing, and management literature. General coverage includes business, economics, management, education and training, forecasting, industrial relations, materials, occupational safety and health, pollution, production, and testing.
Fluidex: (BHRA the Fluid Engineering Centre)	Contains over 150,000 items, updated monthly covering the international literature on fluid engineering subjects from aerodynamics, seals, pumps to ports and offshore technology. Source material drawn from over 1,000 scientific and technical journals scanned annually as well as reports, conferences, books, and standards.	**Searchable Physics Information Notices (SPIN)**	Contains over 400,000 abstracts, covering approximately one third of the world's physics and astronomy journal literature each year. Includes all journal articles published by the American Institute of Physics, nineteen Soviet journals, selected articles from Chinese journals, as well as most of the other American physics and astronomy journals.
Metals Abstracts (METADEX)	Produced and distributed by Materials Information, a joint service of ASM International and Institute of Metals (London), this database contains approximately 650,000 citations on extractive and physical metallurgy. Major literature sources are journal articles with 200 journals abstracted in their entirety and 1,800 journals reviewed for selected input to the database.	**World Textile Abstracts (WTA)**	Covers science, technology, economics, and technical management of textile and related industries, plus all relevant U.S. and U.K. patents. Operated by the Shirley Institute, the database consists of about 150,000 entries, of which 9,000 are new annually.

**Information
Services in
Mechanical
Engineering
(ISMEC)**

Database is devoted to world-wide coverage of mechanical engineering, production engineering, and engineering management and contains 165,000 records. Over 500 technical journals are screened each year for citations, yielding 15,000 new additions annually.

**Information
Services
for the
Physics and
Engineering
Communities
(INSPEC)**

INSPEC dates from 1969, totals over 2.75 million items and is being added to at the rate of 220,000 per annum. Subject breakdown is approximately 53 percent physics, 28 percent electrical and electronics, and 19 percent computer and control.

**Selected
Water
Resources
Abstracts
(SWRA)**

SWRA is the abstract journal of the Water Resources Scientific Information Center (WRSIC). The database became available in machine-readable form in 1969 and contains over 180,000 items with approximately 10,000 records being added per year. Major literature sources covered include scientific and technical journals (80 percent), Government reports (15 percent), patents (1 percent) and published proceedings (4 percent). SWRA covers all aspects of hydrology, hydraulics, water quality and water quantity when water is used as a resource. SWRA also covers the water-related aspects of life sciences, economic & legal aspects of the characteristics, conservation, control, use or management of water.

**World Surface
Coatings
Abstracts
(WSCA)**

WSCA is a product of the Paint Research Association, Middlesex, England and contains 100,000 records dating from 1976 with 800 new citations added monthly. Subject coverage includes: pigments, synthetic resins, cellulose products, adhesives, corrosion, oils, solvents, varnishes and lacquers, inks, industrial hazards, storage and transport.

**Packaging
Science and
Technology
Abstracts
(PSTA)**

Contains over 18,000 abstracts gathered worldwide since 1982 on packaging economy, packaging science and institutions, packaging material, processing, equipment, packs and packages, transport and storage, testing and stress loading.

All Published Searches listed are available from NTIS at the low cost of $45 each. Searches of other databases carry the designations (API), (Ei), (FSTA), (PTO), (IAA), (SWRA), (PSTA), (LSC), (MC), (EDB), (BHRA), (METADEX), (RAPRA), (OCEANIC), (POLLUTION), (PIRA), (SPIN), (WTA), (CONF), (ISMEC), (COM), (INSPEC), or (WSCA) following the title to indicate the source.

To illustrate the breadth of coverage available, we have reproduced on the following pages one (plastics) of the several hundred subjects available in the *1987 Published Searches* catalog.   The legend below provides an explanation of the listings for each published search.   An asterisk by an entry indicates the availability of earlier searches, and a bullet or boldface dot denotes a new listing.

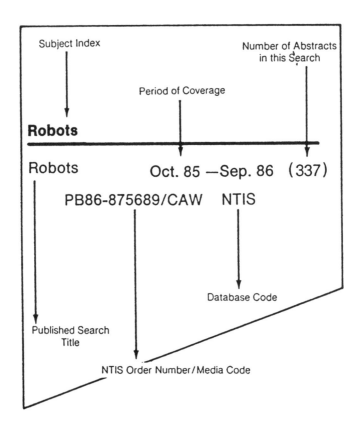

Plastics

- 1980-Aug 86 (169)
  PB86-874880/CAW   WSC

  *Acrylic Adhesives Sep 83-Oct 85 (207)
  PB85-870863/CAW   RAPRA

- Acrylic Coatings: Anticorrosive and Antifouling
  1980-Mar 86 (148)
  PB86-861598/CAW   WSC

- *Acrylic Resins: Acrylates, Acrylamides, and
  General Studies Aug 85-Jul 86 (210) Excludes
  acrylonitrile and methacrylate polymers
  PB86-870847/CAW   Ei

- Acrylic Resins: Methacrylate Polymers 1970-Mar
  86 (300) Covers fabrication, polymerization, and
  properties
  PB86-864188/CAW   NTIS

- Adipate Compounds 1973-Jan 86 (136)
  PB86-857117/CAW   RAPRA

- Amorphous Polymeric Films 1973-Nov 85 (216)
  PB86-852258/CAW   RAPRA

  Anion Exchange Resins: Structure, Formulation,
  and Applications 1977-Jul 85 (101)
  PB85-865384/CAW   RAPRA

  Antioxidants and Stabilizers for Plastics Aug 83-
  Oct 85 (163)
  PB85-870665/CAW   Ei

- Aramid Fibers 1970-Nov 86 (108)
  PB87-851838/CAW   PTO

- *Aramid Fibers: Reinforcement for Plastics and
  Elastomers 1985-1986 (310)
  PB87-852539/CAW   RAPRA

- Aromatic Polyimides 1970-Sep 86 (93)
  PB86-875366/CAW   PTO

- Automobile Coatings: Acrylic and Alkyd Topcoats
  1980-Mar 86 (181)
  PB86-861630/CAW   WSC

- Automobile Coatings: Painting of Plastics 1980-
  Mar 86 (78)
  PB86-861671/CAW   WSC

- Automobile Coatings: Polyurethanes 1980-Mar
  86 (63)
  PB86-861655/CAW   WSC

- Barrier Properties of Packaging Materials 1982-
  Apr 86 (114)
  PB86-864147/CAW   PSTA

- Blow Molding Methods and Machines 1970-Jan
  86 (368)
  PB86-857612/CAW   PTO

- Blowing and Foaming Agents for Polymeric
  Foams Feb 83-1986 (330)
  PB87-852711/CAW   RAPRA

- Blown Films 1970-May 86 (57)
  PB86-868502/CAW   PTO

  Boats: Polymers and Polymer Processing 1973-
  Oct 85 (250)
  PB85-870715/CAW   RAPRA

- Boron, Metal, and Aramid Fiber Reinforced
  Plastics 1973-Nov 86 (194)
  PB87-850541/CAW   RAPRA

- Carbon Fiber Reinforced Plastics (Nongraphite):
  Aerospace Applications 1973-May 86 (249)
  PB86-868759/CAW   RAPRA

- Carbon Fiber Reinforced Plastics (Nongraphite):
  Non-Aerospace Applications 1973-May 86 (183)

  PB86-868445/CAW   RAPRA

- Carbon Fiber Reinforced Plastics (Nongraphite):
  Properties and Tests 1973-May 86 (252)
  Excludes applications
  PB86-868742/CAW   RAPRA

  Cathodic Electrocoating 1980-Apr 85 (267)
  PB85-858462/CAW   WSC

  Cellular Plastic Materials 1970-May 85 (112)
  PB85-860724/CAW   PTO

- Cellulose Membrane Products 1970-Feb 86
  (137)
  PB86-858321/CAW   PTO

- Coextrusion of Plastics 1970-Nov 85 (142)
  PB86-852357/CAW   PTO

- *Combustion of Plastics and Elastomers Nov 80-
  Nov 85 (235)
  PB86-852035/CAW   NTIS

- Computer Aided Plastic and Rubber
  Manufacturing 1973-Jun 86 (293)
  PB86-869807/CAW   RAPRA

- Concrete Polymer Composites 1977-Oct 86
  (272)
  PB87-852117/CAW   RAPRA

  Credit Cards 1970-Aug 85 (276)
  PB85-866150/CAW   PTO

- Crosslinked Polymers 1970-Jul 86 (211)
  PB86-871001/CAW   PTO

- *Electrically Conductive Plastics Aug 83-Apr 86
  (118)
  PB86-867157/CAW   EDB

- *Electrically Conductive Plastics Sep 83-Feb 86
  (190)
  PB86-858917/CAW   Ei

- Electrically Conductive Polymers 1970-Mar 86
  (131)
  PB86-862513/CAW   PTO

- Electron Beam Curing of Polymers 1977-1986
  (269)
  PB87-851978/CAW   RAPRA

- *Engineering Plastics: Properties, Processing, and
  Applications Jan 85-Dec 85 (197)
  PB86-852951/CAW   RAPRA

- Epoxy and Polyester Coatings 1970-Oct 86
  (158)
  PB87-850244/CAW   PTO

- Epoxy Adhesive Compositions 1970-Feb 86
  (170)
  PB86-859972/CAW   PTO

  Epoxy Coatings (Excluding Erosion) Jun 70-Oct
  85 (338)
  PB85-869766/CAW   Ei

- Epoxy Coatings: Anticorrosive and Antifouling
  1980-Mar 86 (246)
  PB86-861440/CAW   WSC

- Epoxy Coatings: Erosion 1970-Apr 86 (147)
  PB86-867090/CAW   Ei

- Epoxy Encapsulants for Electronics 1970-Oct 86
  (78)
  PB86-877180/CAW   PTO

- Epoxy Resins: Aerospace Applications 1973-
  1985 (208)
  PB86-853843/CAW   RAPRA

  Extrusion of High Density Polyethylenes Jul 84-
  Nov 85 (104)
  PB86-850161/CAW   RAPRA

Extrusion of Polypropylenes 1973-Jul 85 (190)
PB85-865368/CAW   RAPRA

● Extrusion Blow Molding of Thermoplastics 1973-
Apr 86 (248)
PB86-865748/CAW   RAPRA

● Fabric Reinforced Composites 1975-Jun 86
(296)
PB86-870623/CAW   WTA

● Fiber Reinforced Composites (Excluding Epoxies
and Polyesters) 1970-Oct 86 (236)
PB86-876984/CAW   Ei

● Fiber Reinforced Composites: Physical Properties
1970-Nov 86 (198) Excludes fiber reinforced
plastics, fiber reinforced laminates, and glass
fiber reinforcements
PB87-851309/CAW   NTIS

● Fiber Reinforced Composites: Technology and
Evaluation 1970-Nov 86 (321) Excludes fiber
reinforced plastics, fiber reinforced laminates,
and glass fiber reinforcements
PB87-851119/CAW   NTIS

● *Fiber Reinforced Epoxy Composites Dec 85-Nov
86 (88)
PB87-851291/CAW   Ei

Fiber Reinforced Polyester Composites 1972-
Oct 84 (106)
PB85-851038/CAW   Ei

● Fiberglass Blistering 1973-Jan 86 (52)
PB86-855889/CAW   RAPRA

● Fiberglass Reinforced Plastics 1977-Nov 85
(263)
PB86-851805/CAW   NTIS

● *Fillers for Plastics Feb 85-Jan 86 (48)
PB86-857190/CAW   Ei

● Flame Retardant Plastics and Elastomers
(Excluding Foams) Oct 83-Sep 86 (318)
Excludes flame retardant plastic foams.
PB86-875457/CAW   RAPRA

● *Flame Retardant Polymeric Foams:
Manufacturing, Applications, and Hazards Jul 84-
Aug 86 (116)
PB86-872447/CAW   RAPRA

● Flame Retardant Polymers 1970-Nov 86 (380)
PB87-851689/CAW   PTO

● *Foamed Plastics: Polyurethane Foams 1979-Mar
86 (205)
PB86-862182/CAW   NTIS

● *Foamed Plastics: Styrene, Silicone, Epoxy, and
Other Polymeric Foams 1979-May 86 (204)
Excludes urethane and isocyanate polymeric
foams
PB86-868486/CAW   NTIS

● Food Packaging: Odor, Flavor, and
Contamination 1976-Feb 86 (205)
PB86-860350/CAW   PIRA

● *Gas Permeability of Polymers 1985-1986 (177)
Water vapor and water permeability of polymers
are included in other bibliographies
PB87-853081/CAW   RAPRA

Glass Fiber Reinforced Plastics: Building and
Agricultural Applications 1975-Oct 85 (133) This
report has been updated and split into 2
seperate titles - one for building applications, the
oth er for agricultural applications
PB85-870681/CAW   RAPRA

Glass Fiber Reinforced Plastics: Chemically
Resistant Applications 1977-Oct 85 (109)
PB85-870475/CAW   RAPRA

● Glass Fiber Reinforced Plastics: Fracture
Mechanics 1975-Jul 86 (293)
PB86-871282/CAW   INSPEC

● Graphite Fiber Reinforced Epoxy 1973-Oct 86
(373)
PB86-877305/CAW   RAPRA

● Heat Resistant Plastics Excluding Nitrogen
Compounds Sep 80-Sep 86 (180)
PB86-875762/CAW   NTIS

● Heat Resistant Polymers 1970-Nov 86 (79)
PB87-850616/CAW   PTO

● Heat Treatment of Thermoplastics and
Thermosets 1973-Jul 86 (271)
PB86-870748/CAW   RAPRA

● High Strength Polymer Composites 1970-May 86
(158)
PB86-869526/CAW   PTO

● Hot Stamping of Plastics 1973-Apr 86 (130)
PB86-868205/CAW   RAPRA

● *Hydrophobic Polymers Mar 83-May 86 (214)
PB86-869450/CAW   RAPRA

● Impact Modifiers for Polymers 1970-Nov 86
(137)
PB87-850624/CAW   PTO

Inflatable Structures and Equipment 1977-Nov
85 (212)
PB86-850229/CAW   RAPRA

*Injection Molding of Polypropylenes Jul 82-Jun
85 (178)
PB85-861508/CAW   RAPRA

● *Injection Molding Machinery and Accessories
Dec 85-Nov 86 (298)
PB87-851507/CAW   RAPRA

● Inorganic Polymers 1976-Oct 86 (148)
PB86-875929/CAW   EDB

● Inorganic Polymers 1970-Nov 86 (290)
PB87-850632/CAW   Ei

● Inorganic Polymers 1970-Apr 86 (159)
PB86-866944/CAW   NTIS

● Interpenetrating Polymer Networks 1970-1986
(124)
PB87-852414/CAW   Ei

● *Metallization of Plastics Mar 83-1986 (118)
PB87-852182/CAW   Ei

● *Metallization of Plastics 1984-Nov 86 (97)
Excludes immersion plating and ion plating
technology
PB87-850764/CAW   METADEX

● *Metallization of Plastics Nov 85-Nov 86 (95)
PB87-851812/CAW   RAPRA

● Metallizing Thermoplastics and Thermosets
1973-1985 (347)
PB86-853181/CAW   RAPRA

● *Mineral Fillers in Plastics and Elastomers Jun
83-May 86 (272)
PB86-868577/CAW   RAPRA

Mold Release Agents 1970-Sep 85 (112)
PB85-867497/CAW   PTO

● *Mold Release Agents for Rubbers and Plastics
1985-1986 (95)
PB87-852125/CAW   RAPRA

- Molding Thermosetting and Thermoplastic
  Structural Foam 1970-Oct 86 (271)
  PB86-876794/CAW   Ei
- *Molding Thermosetting and Thermoplastic
  Structural Foam Dec 84-1985 (54)
  PB86-854510/CAW   RAPRA
- Multilayer Plastic Films 1970-Aug 86 (121)
  PB86-873379/CAW   PTO
- Nitrogen Containing Heat Resistant Plastics
  1976-Aug 86 (261)
  PB86-873866/CAW   NTIS

  Nondestructive Testing of Plastics 1970-May 85
  (279)
  PB85-860781/CAW   NTIS
- *Nondestructive Testing: Rubbers and Plastics
  Nov 83-Oct 85 (128)
  PB85-870061/CAW   RAPRA
- Oriented Films in Packaging 1982-Jun 86 (55)
  Emphasizes food packaging applications
  PB86-871159/CAW   PSTA
- Pharmaceutical Packaging: Plastics and
  Elastomers 1973-Nov 86 (363)
  PB87-850947/CAW   RAPRA
- Photochromic Polymers 1973-Sep 86 (77)
  PB86-875853/CAW   RAPRA
- Photoconductive Plastics 1975-Jan 86 (254)
  PB86-856127/CAW   INSPEC
- Piezoelectric and Pyroelectric Polymers
  (Excluding Vinylidene Fluoride Polymers 1975-
  Oct 86 (237)
  PB86-875945/CAW   INSPEC
- *Plastic and Rubber Materials used in Athletic
  Equipment 1983-May 86 (237)
  PB86-870003/CAW   RAPRA
- *Plastic Bottles Used in the Food Industry Dec
  85-Nov 86 (47)
  PB87-850996/CAW   FSTA

  Plastic Building Materials for Underground
  Construction 1977-Oct 85 (304)
  PB85-869634/CAW   RAPRA
- Plastic Coatings 1970-Feb 86 (123)
  PB86-857547/CAW   PTO
- Plastic Electrostatic Coatings 1973-May 86
  (185)
  PB86-870300/CAW   RAPRA
- Plastic Explosive PETN: Penaerethritol
  Tetranitrate 1970-Jun 86 (127)
  PB86-870573/CAW   NTIS
- Plastic Explosive RDX:
  Cyclotrimethylenetrinitramine 1970-May 86 (262)
  PB86-870268/CAW   NTIS
- Plastic Films in Packaging 1982-1986 (169)
  PB87-852380/CAW   PSTA
- Plastic Foam Coating: Processes, Formulations,
  and Applications 1975-Apr 86 (299)
  PB86-864410/CAW   RAPRA
- Plastic Fuel Tanks in Automobiles and Aircraft
  1973-1986 (247)
  PB87-852570/CAW   RAPRA
- Plastic Lenses: Fabrication and Applications
  1970-Mar 86 (100)
  PB86-859386/CAW   PTO
- Plastic Lenses: Fabrication and Applications
  1977-1986 (303)

PB87-853073/CAW   RAPRA

Plastic Sheathing and Siding: Building
Applications 1978-Jun 85 (194)
PB85-863165/CAW   RAPRA
- *Plastic Window Frames Mar 83-1985 (179)
  PB86-853835/CAW   RAPRA
- Plastic Window Materials: Frames and Glazing
  1977-1985 (276)
  PB86-853983/CAW   RAPRA
  *Plastic-Based Flooring Materials Jun 83-Nov 85
  (151)
  PB85-872398/CAW   RAPRA
- Plastics and Elastomers as Moisture Barriers
  1973-Nov 86 (225)
  PB87-850491/CAW   RAPRA
- *Plastics and Elastomers as Protective Coatings
  Feb 85-Jan 86 (61)
  PB86-856549/CAW · RAPRA

  Plastics and Elastomers: Dental Applications
  1973-Oct 85 (237)
  PB85-870491/CAW   RAPRA
- *Plastics and Elastomers: Electrical Conductivity
  Jan 86-Dec 86 (149) Excludes electromagnetic
  shielding applications of conductive plastics
  PB87-852653/CAW   RAPRA
- Plastics and Elastomers: Electron Beam
  Irradiation Effects 1977-Oct 86 (286)
  PB86-876166/CAW   RAPRA
- *Plastics and Elastomers: Impact Strength
  Additives Jul 84-Aug 86 (110)
  PB86-872439/CAW   RAPRA
- *Plastics and Elastomers: Machining Processes
  and Properties May 85-Mar 86 (55) Excludes
  molding operations
  PB86-862810/CAW   RAPRA
- Plastics and Elastomers: Magnetic Properties
  1973-1985 (148)
  PB86-853819/CAW   RAPRA
- Plastics and Elastomers: Mechanical Degradation
  1973-Feb 86 (186)
  PB86-860954/CAW   RAPRA
- Plastics and Elastomers: Military Applications
  1977-1986 (300)
  PB87-853099/CAW   RAPRA

  Plastics and Elastomers: Mining Applications
  1973-Nov 85 (267)
  PB85-872356/CAW   RAPRA
  *Plastics and Elastomers: Nondestructive Testing
  Jul 83-Nov 85 (177)
  PB86-850211/CAW   RAPRA

  Plastics and Elastomers: Offshore Applications
  1973-Nov 85 (320)
  PB85-872406/CAW   RAPRA
- Plastics and Elastomers: Ozone Degradation
  1973-1986 (187)
  PB87-852604/CAW   RAPRA
- *Plastics and Elastomers: Pharmaceutical
  Applications (Excluding Packaging) Apr 83-Feb
  86 (169)
  PB86-858909/CAW   RAPRA

  Plastics and Elastomers: Plating Methods ¡1973-
  Nov 85 (262)
  PB86-850187/CAW   RAPRA
- Plastics and Elastomers: Radiation Crosslinking

1973-Mar 86 (342)
PB86-862539/CAW   RAPRA

● Plastics and Elastomers: Space Applications
1977-1985 (234)
PB86-855046/CAW   RAPRA

● Plastics and Rubbers: Horticultural Applications
1978-1985 (237)
PB86-856101/CAW   RAPRA

● Plastics for Automotive Bumpers and Headlights
1979-Oct 85 (330)
PB86-876877/CAW   RAPRA

● Plastics for Automotive Bumpers and Headlamps
Nov 85-Oct 86 (116)
PB86-876885/CAW   RAPRA

  *Plastics Decomposition Oct 83-Sep 85 (110)
PB85-867919/CAW   Ei

  *Plastics Extrusion Mar 83-Mar 85 (196)
PB85-856490/CAW   Ei

● Plastics Used as Building or Construction
Materials 1975-Apr 86 (268)
PB86-863008/CAW   Ei

● Plastics: Space Applications 1972-Aug 86 (196)'
PB86-873452/CAW   IAA

Plastisols 1970-Oct 85 (125)
PB85-871309/CAW   Ei

Plastisols 1970-1984 (139)
PB85-851392/CAW   PTO

Plastisols 1980-Nov 85 (114)
PB85-871648/CAW   WSC

● Plastisols: Formulations and Applications 1973-
Apr 86 (336)
PB86-864451/CAW   RAPRA

● *Plating on Plastics and Elastomers Nov 83-1985
(309)
PB86-857133/CAW   Ei

● Polycarbonate Compositions 1970-May 86 (79)
Excludes flame retardant compositions
PB86-868981/CAW   PTO

● Polycarbonates: Optical Properties and
Applications 1973-1986 (276)
PB87-852455/CAW   RAPRA

● *Polycarbonates: Properties, Synthesis, and
Applications 1977-Mar 86 (246)
PB86-860590/CAW   NTIS

Polyester and Epoxy Resins: Abrasion
Resistance 1977-Oct 85 (256)
PB85-869907/CAW   RAPRA

● Polyester Polyols 1970-Sep 85 (117)
PB86-875754/CAW   PTO

● Polyether Polyols 1970-Aug 86 (117)
PB86-872793/CAW   PTO

● Polyether-Etherketon Materials 1973-Apr 86
(120)
PB86-863982/CAW   RAPRA

Polyethylene Terephthalate Packaging Materials
1982-Nov 85 (121)
PB86-851094/CAW   PSTA

● Polyethylenes: Light stability 1973-1985 (131)
PB86-853355/CAW   RAPRA

● Polymer Powder Coatings: Industrial and Building
Application: 1973-Feb 86 (115)
PB86-857588/CAW   RAPRA

● Polymer Radiation Curing: Epoxies, Phenolics,

Fluorocarbons, and Silicones 1970-Jan 86 (255)
PB86-857752/CAW   NTIS

● Polymer Radiation Curing: Polyolefins and
Acrylics 1970-1986 (216)
PB87-852463/CAW   NTIS

● Polymer Radiation Curing: Styrenes and Vinyls
1970-Apr 86 (173)
PB86-864584/CAW   NTIS

● Polymeric Foam Mattresses 1977-Aug 86 (206)
PB86-872835/CAW   RAPRA

● *Polymeric Gels and Hydrogels Dec 85-Nov 86
(67)
PB87-851242/CAW   Ei

● Polymeric Materials: Decomposition and
Corrosion Jun 74-Aug 86 (299)
PB86-873593/CAW   IAA

● *Polymeric Prosthetic Devices Jul 84-Aug 86
(168)
PB86-872454/CAW   RAPRA

● *Polymeric Roofing Materials Jul 84-Nov 86 (209)
PB87-851317/CAW   RAPRA

● *Polymeric Sealants Jan 85-Dec 85 (74)
PB86-855855/CAW   RAPRA

● Polymers and Copolymers: Glass Transition
Temperature 1971-Mar 86 (267)
PB86-862695/CAW   Ei

● Polymers in Vibration Damping and
Soundproofing 1970-Feb 86 (100)
PB86-858966/CAW   PTO

● Polyphosphazenes Dec 73-Nov 86 (253)
PB87-851358/CAW   RAPRA

● Polyurethane Adhesives 1980-Jan 86 (133)
PB86-855954/CAW   WSC

● *Polyurethane Resins: Synthesis and Properties
1979-1986 (305) Excludes polyurethane foams
PB87-852703/CAW   NTIS

  *Polyurethane/Polyisocyanurate Foam Thermal
Insulation Nov 83-Oct 85 (136)
PB85-869329/CAW   RAPRA

● Polyurethanes: Curing 1980-Nov 86 (220)
PB87-850087/CAW   WSC

● Polyurethanes: Light Stability 1973-1986 (96)
PB87-852372/CAW   RAPRA

  *Polyvinylidene Fluoride Nov 83-Oct 85 (99)
PB85-870194/CAW   Ei

Polyvinylidene Fluoride 1970-Nov 85 (193)
PB85-871622/CAW   NTIS

● Polyvinylidene Fluoride 1970-Apr 86 (123)
PB86-867355/CAW   SPIN

● Polyvinylidene Fluoride: Reinforcers and
Processing 1977-Apr 86 (174)
PB86-863768/CAW   RAPRA

Polyvinylidene Fluoride: Structure and
Degradation 1976-Oct 85 (241)
PB85-869618/CAW   RAPRA

● Prepregs 1970-Feb 86 (110)
PB86-858743/CAW   PTO

● Prepregs: Applications 1973-1986 (195)
PB87-852596/CAW   RAPRA

● Prepregs: Composition and Processing 1973-
1986 (319)
PB87-852588/CAW   RAPRA

● Prosthetic Devices: Polymeric Materials
Utilization 1970-Nov 86 (231)

PB87-851440/CAW   Ei

Pultrusion 1970-Oct 85 (71)
PB85-869188/CAW   Ei

● Pultrusion of Reinforced Plastics 1973-Oct 86 (302)
PB86-876653/CAW   RAPRA

● PVC Plastics: Light Stability 1973-Nov 85 (148)
PB86-852175/CAW   RAPRA

● Radiation Curable Polymer Coatings 1970-Jun 86 (139)
PB86-860509/CAW   PTO

● *Radiation Curing of Polymers Dec 83-Nov 86 (324)
PB87-851457/CAW   Ei

● *Reaction Injection Molding: Polyamide and Polyurethane Resins 1973-Mar 86 (277)
PB86-863420/CAW   RAPRA

● Reaction Injection Molding: Polyamide and Polyurethane Resins (Applications) 1973-Mar 86 (260)
PB86-862224/CAW   RAPRA

● Recycling Plastic Scrap: Injection Molding 1973-May 86 (58)
PB86-669484/CAW   RAPRA

● Reinforced Plastics: Electrical Applications 1973-Feb 86 (274)
PB86-859964/CAW   RAPRA

● Reverse Osmosis Resins 1975-Mar 86 (186)
PB86-859030/CAW   RAPRA

Rubber and Plastics Manufacture: Use of Wax Additives 1973-Nov 85 (240)
PB85-872372/CAW   RAPRA

Seals and Sealants: Physical Stability 1973-Oct 85 (269)
PB85-870616/CAW   RAPRA

● Sheet Molding Compounds 1970-Apr 86 (97)
PB86-866373/CAW   PTO

● Sheet Molding Compounds: Automotive Applications 1973-Jul 86 (212)
PB86-873163/CAW   RAPRA

● *Silicone Resins: Chemistry 1977-Sep 86 (141)
PB86-875721/CAW   NTIS

● Solvent Free Polymeric Coatings 1973-Nov 86 (210) Includes protective, adhesive, and insulative coatings and sealants
PB87-852638/CAW   RAPRA

● Solventless Polymers 1970-May 86 (167)
PB86-868569/CAW   PTO

● Stretchwrap and Shrinkwrap Films 1973-Apr 86 (196)
PB86-864469/CAW   RAPRA

● Structural Foam Applications 1973-Apr 86 (262)
PB86-865755/CAW   RAPRA

● Structural Foam: Finishing and Decorating 1977-Jan 86 (159)
PB86-860608/CAW   RAPRA

Thermoforming 1970-Nov 84 (184)
PB85-851343/CAW   PTO

● Thermoplastic Interpenetrating Polymer Networks 1973-Jan 86 (54)
PB86-856721/CAW   RAPRA

Thermoplastic Laminates and Foams 1970-Sep 85 (92)
PB85-868594/CAW   PTO

● Thermoplastic Packaging 1982-Mar 86 (170)

PB86-863206/CAW   PSTA

● Thermoplastic Polyesters 1970-1986 (204) Excludes foams and laminate materials
PB87-852174/CAW   PTO

Thermoplastics and Elastomers: Melt Flow Indices 1973-Mar 85 (239)
PB85-857530/CAW   RAPRA

● Thixotropy 1973-Apr 86 (214)
PB86-864501/CAW   RAPRA

● Toxicity: Polymeric Materials in Food-Contact Applications 1977-Oct 86 (237)
PB86-875960/CAW   RAPRA

● Toy Manufacturing: Plastic and Rubber Materials 1973-Feb 86 (240)
PB86-860947/CAW   RAPRA

Twin-Screw Extruders 1973-Nov 85 (260)
PB85-872307/CAW   RAPRA

● Ultra-High Molecular Weight Polyethylene 1973-Feb 86 (167)
PB86-857927/CAW   RAPRA

● Ultra-High Molecular Weight Polymers 1970-Jan 86 (59)
PB86-856622/CAW   Ei

Ultrasonic Welding of Plastics 1973-Nov 85 (195)
PB85-872281/CAW   RAPRA

● Urea Formaldehyde in Building Materials 1973-Nov 86 (320)
PB87-851952/CAW   RAPRA

● Urea/Phenol/Melamine Formaldehyde Polymeric Resins 1970-Apr 86 (151)
PB86-864477/CAW   NTIS

● Urea/Phenol/Melamine Formaldenyde Polymeric Resins 1970-Nov 85 (265)
PB86-851235/CAW   Ei

● Urethane Coatings Jun 70-Nov 86 (320)
PB87-851200/CAW   Ei

Vacuum Forming of Plastics 1973-Jun 85 (203)
PB85-863488/CAW   RAPRA

● Vinyl Chloride and Polyvinyl Chloride: Toxicology 1978-1985 (231)
PB86-856119/CAW   LSC

● *Vinylidene Fluoride Polymers and Copolymers: Electrical Properties and Applications Apr 85-Feb 86 (60) Emphasis on piezoelectric and pyroelectric properties
PB86-859832/CAW   INSPEC

● Water Permeability of Coatings 1980-Feb 86 (105)
PB86-859758/CAW   WSC

● Water Permeability of Polymers 1973-Sep 86 (141)
PB86-875127/CAW   RAPRA

● Water Vapor Permeability of Polymers (Excluding Gas Permeability) 1973-Nov 86 (305)
PB87-851465/CAW   RAPRA

● Water-Based Polyester Coatings 1980-Aug 86 (139)
PB86-874872/CAW   WSC

● Welding of Plastics: Excluding Ultrasonic Welding 1973-Jul 86 (385)
PB86-872173/CAW   RAPRA

● X Ray Diffraction Studies: Polymers 1970-Jul 86 (135)
PB86-873098/CAW   NTIS